The War of the Copper Kings

GREED, POWER, AND POLITICS:
THE BILLION-DOLLAR BATTLE FOR BUTTE,
MONTANA, THE RICHEST HILL ON EARTH

By C. B. Glasscock

Ignorance plays the chief part among men, and the multitude of words, but opportunity will prevail.
— Cleobulus

RIVERBEND
PUBLISHING

ISBN 978-1-931832-21-2

Published by Riverbend Publishing, Helena, Montana

Cover design by DD Dowden
Text design by Teresa Record

RIVERBEND PUBLISHING
P.O. Box 5833
Helena, MT 59604
1-866-787-2363
www.riverbendpublishing.com

To

THE OLD DEALERS

In sympathy with their errors,
and admiration of their
accomplishments

ACKNOWLEDGMENTS

For the material facts and expressions of varied personal opinion contained in this narrative the author acknowledges his indebtedness to many persons and printed sources. First among them, perhaps, should be named the Historical Society of Montana at Helena, Mr. David Hilger, librarian, and his very able and courteous assistant, Mrs. Anne McDonnell. They opened to research the files of scores of Montana newspapers of the decade from 1895 to 1905 which are quoted in the text.

Men and women who contributed generously in personal interviews number scores. Of those who took prominent parts in the great drama and who told the writer of their experiences and reactions, the principals included John J. McHatton, a judge in Butte for many years, and one of F. Augustus Heinze's leading attorneys; William Scallon, long the resident director of the Amalgamated Copper Company's war against Heinze in Butte; Charles R. Leonard, Heinze's personal counsel in the final settlement with the Amalgamated; Lee Mantle, once United States Senator from Montana and a prominent citizen, mine owner and publisher in Butte for many successful years; John Gillie, for half a century a mining engineer in Montana, for many years chief engineer of the Amalgamated, personal friend of Marcus Daly, and at the time of this writing still consulting engineer of the Anaconda Copper Mining Company. John MacGinniss, Heinze's right hand man in much of the action, asserted that if the men named had told their stories, nothing more was needed to complete the picture.

Completion however was sought in interviews and other assistance generously given by James W. Scott, veteran newspaper reporter who covered the Pennsylvania vs. Rarus suit in 1899 and many others of the scores of historic legal actions; Miss Horgan, Miss Sternfels and Miss O'Meara of the Butte Public Library; Samuel

Barker and Arthur V. Corry, mining engineers; Judge George M. Bourquin, Judge W. I. Lippincott, D. J. Charles, Mrs. T. J. Murray, Jerry Clifford and various residents and pioneers of Butte in all walks of life who helped to reveal the varied life of the town as well as the economic and political drama. Alex Leggat gave free access to his private library with its superior collection of Montaniana. Lew. L. Callaway, Chief Justice of the Supreme Court of Montana, and noted authority on the early history of the territory, contributed generously of material on the vigilante days.

Outside of these sources, former residents of Butte now living in other cities were equally kind and helpful. Those interviewed included Mrs. E. B. Weirick, president of the Montana Woman's Club of Southern California; Charles Henderson, once sheriff of Silver Bow County; Ben E. Stack, whose memories of Butte date back to 1876; Lee Hayes, mining engineer of Butte as far back as 1897; Dr. W. L. Renick; Will L. Clark, Clerk of the District Court in 1888, and later employed at various times by Marcus Daly, William A. Clark and F. A. Heinze; Charles T. Shearer, an active newspaper man of the early days; Judge Sydney Sanner, Mrs. Nell A. DeLeon, and others. Mrs. Isaac Edinger introduced the writer to numerous productive sources of material. "Ike" Edinger contributed from his memories of more than fifty years as rancher, bank director and public-spirited citizen of Montana.

The W. A. Clark Memorial Library in Los Angeles, under the direction of Miss Cora E. Sanders, opened its files and stacks to free research.

Printed sources of information, in addition to the Montana newspaper files, include various magazines which recognized the drama twenty-odd years ago as a matter of national interest and moment: *Everybody's, McClure's, Current Literature, Donahoe's, Leslie's, Engineering & Mining Journal, Cosmopolitan.*

Books which have afforded both facts and sidelights on the period and activities involved include Thomas W. Lawson's

Frenzied Finance, Ridgway-Thayer Co., 1905; Jerre C. Murphy's *The Comical History of Montana,* E. L. Scofield, 1912; Tom Stout's *Montana,* American Historical Society, 1921; Helen F. Sanders' *History of Montana,* Lewis Publishing Co., 1913; George Wesley Davis' *Sketches of Butte,* The Cornhill Co., 1921; *Dictionary of American Biography; U. S. Senate Reports,* 56th Congress, 1st session, 1899-1900; T. J. Dimsdale's *Vigilantes of Montana,* 1866; Hoffman Birney's *Vigilantes,* Penn Publishing Co., 1929.

The extent and variety of the sources acknowledged should be emphasized because of the highly controversial nature of the subject. The violence of the warfare was demonstrated most vividly in the course of research by the fact that numerous persons interviewed still maintained an intense partisanship a third of a century after the action. For that reason the names and comments of some of the old-timers may open old wounds and stir indignation anew, but at the same time should be convincing evidence that an honest effort has been made to narrate nothing more than essential facts and opinions. Those who feel that their own particular hero has not been revealed in the most favorable light can only be referred to the names and standing of witnesses for the opposition.

C. B. GLASSCOCK

CONTENTS

The War of the Copper Kings

CHAPTER I

CIVILIZATION AT SCRATCH

A GROUP of ragged, bearded, weary men, fleeing from hostile Indians, made camp beside an alder-lined creek high in the Rocky Mountains in the spring of 1863. One washed two dollars and forty cents' worth of gold from a panful of sand. Unsuspected in that sand was the promise of one hundred million dollars in gold—a promise soon to be realized.

Latent in the surrounding ground were the makings of more than one hundred murders, more than a score of hangings without process of law, more than two billion dollars in material wealth, a corporation designed to rule the electric light and power of the world, battles of giants in the depths of the earth and in the capital of the nation, and a financial panic that staggered the country in 1907.

Bill Fairweather and his associates were on their way to the town of Bannack, the first rich gold camp in all the vast region of the northern Rockies, opened only the year before by John White. They had been prospecting around the headwaters of Yellowstone River, and had escaped with their scalps only because Fairweather, six feet two of bone and sinew, with a mustache long enough to loop behind his ears, and a red beard halfway to his gun belt, had bluffed the Sioux medicine men to a standstill. There is a legend still accepted in Alder Gulch and by some students of the early history that Bill sustained that bluff by juggling two writhing, hissing, striking rattlesnakes before the shrinking noses of the Injuns.

Harry Edgar, one of the party, who left a detailed diary telling of that historic journey and discovery, wrote that Fairweather pulled up the "medicine bush" in the center of the lodge where they were being put through an ordeal preliminary to death, and whacked the medicine man across the face with it. It seems likely that if Bill had juggled snakes as well as the sacred bush, Edgar's diary, meticulous in names, dates and incidents, would not have overlooked that dramatic detail.

However, it is of minor importance except to indicate the character of the man, and the status of the civilization in which he lived. Whatever the reason, the prospectors were freed with the understanding that they would return forthwith to Bannack and go no farther into Sioux territory.

But the prospecting trip had not been a success. Food was almost exhausted, and the horses were footsore and worn by the time the party reached the western slope of the Tobacco Root Range, seventy-odd miles from Bannack. Camped in a narrow canyon, Bill Fairweather and Harry Edgar were detailed to tend the horses while the four others traveled up the creek to prospect. Fairweather, looking for grass on which to stake the horses, noted a rock formation which suggested gold.

At the moment the thing he most wanted was tobacco, and he told Edgar there might be tobacco money in the rock. While Edgar carried the first pan of broken rock to wash in the creek, Fairweather picked up a nugget which subsequently weighed in at two dollars and forty cents. Edgar's pan produced the same amount in fine gold.

"Four dollars and eighty cents!" Edgar's diary exclaims. "Pretty good for tobacco money. We went and got another pan and Bill panned that and got more than I had; I got the third one and panned that—best of the three, good enough to sleep on. We came to camp, dried and weighed our gold, and altogether there was twelve dollars and thirty cents."

Ambition promptly soared above tobacco money to a grubstake of one hundred and fifty dollars. In two days the takings of the six men reached one hundred and eighty-nine dollars, but their grub was too low to permit a longer stay. They staked their claims, named the ravine Alder Gulch, agreed to keep the secret until they could come back and get the best of it, and started for Bannack, making their own trail.

Weariness, hunger, hardship and danger were forgotten. This was no group of failures plodding over the hill and down into the town of Bannack, feebly giving thanks to God for escape from the tortures and scalping knives of the Sioux. "The raggedest lot that ever was seen," says Harry Edgar's diary, "but happy." And Bannack, made up of three or four hundred men of similar breed and background, was quick to note and interpret that happiness.

"Friends on every side," Edgar's record reveals. "Bob Dempsey grabbed our horses and cared for them. Frank Ruff got us to his cabin. Salt Lake eggs, ham, potatoes, and everything. Such a supper! . . After I got my store clothes on I was sitting in a saloon talking with some friends. There were lots of men there who were strangers to me. They were telling that we brought in a horse-load of gold—and not one of the party had told that we even found color. Such is life in the Far West."

Such indeed has been the life and character of all true prospectors since men began to abandon pick-and-pan profits of one hundred dollars a day each in the California diggings in 1849 to stampede over perilous mountains in search of a rumored "source of gold."

Within a day the entire mining district around Bannack was aflame with the excitement of a vast new gold discovery, although only the Fairweather party had the slightest idea of where it might be, and they were maintaining strict secrecy. When, newly outfitted and supplied with food, the party left Bannack two days later and headed toward the Beaverhead River, nearly half the town was at their heels, and others were waiting in every gulch to join the rush, they knew not whither. Two hundred men, on foot and horseback, trailed the discoverers, camping all around them, watching through every hour of day and night to prevent escape and the secret looting of the unknown claims.

Wise in experience of the day, the wilderness, and the character of men who dogged their steps, the Fairweather party called a halt a day before they neared their goal. There in a barren valley they called a miners' meeting to draw up laws and regulations of the new district and assure themselves of the inviolability of the claims they had staked. Without such rules and pledges they declared they would not go on. The rules were adopted and the pledges given. The discovers took up the journey, with the eager crowd at their heels.

"When we came to the creek," says Edgar's journal, "and were going up I said to them: 'This is the creek.' Such a stampede! I never saw anything like it before. Got to Discovery about 4 o'clock. The creek is all staked." And that creek was more than ten miles long.

In a week there was a town of half a thousand men, a town of tents, brush huts and shacks in the neighborhood of Bill Fairweather's discovery. Some erstwhile resident of the seceded South who preferred to register his principles with pen rather than sword, as he might have been doing that week with General Lee moving from Fredericksburg in his second invasion of the North, named the camp Varina, in honor of the wife of Jefferson Davis, President of the Confederacy. A week later Dr. G. G. Bissell, elected first president of the district by a miners' meeting, scratched the name off papers submitted for his signature and substituted Virginia City, announcing that he would be damned if he would sign any papers in a town named for the wife of the rebel President.

Virginia City it was, and in ruins still is; to sad confusion with the Virginia City of the Comstock Lode. In ninety days ten thousand men, and a sprinkling of women for their entertainment, flocked in. The whole area of what is now Montana could not supply so many. They came laboriously by steamboat and barge up the Mississippi River and over two hundred miles of perilous trails from Fort Benton, by foot and horseback and freight wagon from Salt Lake City, far to the southward, by any means of transportation from the declining placers of California and even from the riches of the Comstock in Nevada. Alder Gulch was only twelve or fifteen miles long. It could hardly hold its residents. Virginia City could not, and quickly gave up trying. Half a dozen camps, each boasting a name and a miners' government, sprang into being along the few miles of meandering wagon trail below and above Virginia City. Nevada City was the first. Adobetown, Central, Pine Grove, Highland, Summit and Junction were not far behind nor far away.

It was, in effect, one long town with a single crooked street. That street was bordered from end to end with such an assortment of saloons, gambling houses, dance-halls, cribs, stores, stables, blacksmith shops and miscellaneous businesses as a more civilized world is unlikely to see again. Perched on the sides of the gulch above were the cabins, tents and huts of the thousands of miners who were washing out a million dollars worth of gold each month and scattering it across the bar counters and counterpanes, to make their town a place of fame and infamy. A rich town, a wild town, a wicked town.

Bannack, on the site of the year-old Grasshopper Creek diggings, seventy-odd miles to the westward, was the only other town worth mentioning in the five hundred miles of mountainous wilderness north and south between Salt Lake City and the Missouri steamboat landing at Fort Benton. From east to west there was not a settlement one-tenth as populous between Minneapolis and Portland, Oregon. Portland itself boasted only three thousand population while Alder Gulch was roaring with the shouts of ten thousand. There was only one other mining camp then in the world to equal it—Virginia City, Nevada, nearly a thousand miles away by unmapped trails over mountains and desert.

Understanding of the vast extent of this area, almost unpopulated except by savages outside the two settlements of Alder Gulch and Bannack, is essential for understanding of the nature of life therein, and the crudities of the civilization of that day and place.

Into that area, on the heels of the placer miners, first lured by swift rumors of the rich diggings of Grasshopper Creek, and increased to a tidal wave of adventurous humanity by the waiting wealth of Alder Gulch, came a motley throng. Bannack was only a few months old when the news of Alder Gulch spread with electric speed throughout the West, already attuned to such news by the California gold rush and the first bonanza of the Comstock. Mountain men and desert men, the bold women of the dance-halls and cribs in St. Louis, Chicago and San Francisco, gamblers and desperadoes rushed in for the kill.

In the very week that Bill Fairweather and his five companions discovered gold in Alder Gulch, one Henry

Plummer had been elected sheriff of the Bannack district. Plummer was a smooth shrewd frontiersman with the appearance of a gentleman, the charm of Satan, the organizing ability of an Al Capone and the moral equipment of a John Dillinger. He was the first truly notable gangster of America, though no one suspected it. The residents of Bannack knew only that he was quick and accurate with a gun, apparently courageous, neat and clean in dress, polite and charming to women, affable with men.

They did not know that he had been marshal of the pioneer mining town of Nevada City, California, six years earlier. They did not know that there he had won the favors of a married woman, killed the husband who surprised him, and started the American gangsters' prostitution of the law by originating a sob-sister trick which promptly won his release. They did not know that he had wounded another man fatally in a subsequent row in a bawdy-house. They did not know that he had killed a third man in a similar quarrel in Nevada City and escaped jail by means of smuggled guns. They did not know that he had organized a road-agent gang to rob miners and express messengers of the gold from the Oro Fino diggings in Idaho and that he had come to Bannack direct from those crimes in 1862 as the secret leader of a group of men who would and did kill without one moment's hesitation, either for a poke of gold or the favors of a woman.

The name and fame of bad-men, murderers and bandits, did not spread with the speed of the rumors of new gold strikes in the West of that day. There was virtually no government between the Missouri River and the Sierra Nevada. There were few and insignificant newspapers. All forms of transportation and communication were slow and arduous. No community could or did worry much about the troubles of another. When

Plummer and a few of his henchmen departed hastily from the Lewiston section of Idaho after the killing of one Patrick Ford in his Oro Fino saloon, both Lewiston and Oro Fino were content. The banditry of the region was checked, and that was enough.

There were no local officers to send after Plummer, no newspapers to spread his picture and his record over a thousand miles of wilderness. Each new camp accepted whatever Fate might send it, and asked no embarrassing questions. Plummer was accepted in Bannack at his face value, which seemed high. Members of his Idaho road-agent gang, including Charley Reeves, Jack Cleveland, Cyrus Skinner and Bill Bunton, were accepted as part of the riffraff that always appeared in the unpoliced frontier camps to run saloons, gamble, drink and carouse.

With more calculating shrewdness than numerous gang leaders of a later date, Plummer concealed his former association with these men. He met them casually as he met the honest miners and business men of the camp, in the saloons. They, following the code of their kind, were equally uncommunicative concerning their past. Plummer's reputation remained unquestioned and unsmirched.

But he had quarreled with one, Jack Cleveland, when the two men were paying court to a country beauty named Electa Bryan. Having won the girl's promise of marriage, Plummer feared Cleveland's possible revelations. To remove that danger, when Cleveland, drunk, started another quarrel in a Bannack saloon, Plummer put half a dozen slugs into his body. Although mortally wounded, Cleveland proved himself the first exponent of a code of silence made famous by scores of dying gangsters of the prohibition era. He declined to explain the cause of his troubles with Plummer, or to reveal Plummer's record. He died with the boast of a modern gangster that he could take it.

Plummer, however, was fearful that the dying man had talked too much to Henry Crawford, first sheriff of Bannack. Therefore Plummer transferred his guilt-inspired enmity to Crawford. Excitement over the killing of Cleveland was heightened, but somewhat diverted from Plummer a few days later when three other bad-men, Bill Moore, Charley Reeves and Bill Mitchell, shot up an Indian camp on the edge of town, killing a white man and three Indians, and wounding several whites. A volunteer posse trailed and trapped the killers, and to their astonishment found Plummer with the group. In the ensuing informal trial, with the friends of the killers openly threatening the jury, a compromise verdict was reached with a sentence of banishment. Plummer was cleared of the murder of Cleveland. The other killers returned when a miners' court, dominated by their friends, revoked the sentence of banishment.

Bannack had gone on record as being without law or safety for life or property, a haven for murderers and robbers. Still there was no evidence of Plummer's business association with the gang. His appearance with the fugitives after the wholesale murders at the Indian camp had required considerable explaining, but he had won the approval of Bannack by being equal to the task. At the same time he had proved himself a smarter person than most of the rough men of the camp.

But he was not convinced that Jack Cleveland had died without revealing to Sheriff Crawford a past in which Plummer had killed at least four men and superintended the robbery of scores. Crawford must be silenced. To a man of Plummer's character, that meant to kill. So Plummer forced a quarrel upon Crawford at every opportunity, and finally drove the sheriff to such desperation that he took a potshot at Plummer from across the street, and tore Plummer's right arm from wrist to elbow with the bullet.

For a time it appeared that the arm must be sacrificed, but the nursing of friends saved arm and life, and before the wounded man was out of bed Crawford had closed his butcher shop and fled the country. Bannack was without a sheriff. Plummer was relieved of the fear of disclosure of his past, and he was elected to the vacant office. The plot was complete for the murder of more than one hundred men, the robbery of scores of individuals, parties and express boxes carrying the gold of the newly discovered Alder Gulch and the year-old Grasshopper Creek diggings. That was the state of civilization in the Rocky Mountains in 1863. Its immediate revelations are worth a chapter to set the stage for the greater robberies and battles of Butte some thirty years later, and the still greater robberies accomplished in Wall Street in the next decade.

CHAPTER II

VIGILANTES SHOW THE WAY

HONORABLE position in the community established by his election to the office of sheriff, Henry Plummer made a dangerous journey by horseback over two hundred miles of mountain trails to Sun river and brought back as a bride the Electa Bryan concerning whom he had quarreled with Jack Cleveland, to the latter's death. The marriage lasted but ten weeks. Plummer perhaps reverted to type in his domestic life before he did in professional life. In any event, the bride departed for Salt Lake City and was never seen again in the Montana country.

In the meantime, the new sheriff had appointed four deputies, of whom only one, D. H. Dillingham, appears to have been an honest and conscientious man. The three others, Ned Ray, Buck Stinson and Jack Gallagher, were murderous desperadoes of the most vicious type, and generally suspected of that character. How Plummer could have appointed them without subjecting himself to grave suspicion is a mystery unsolved. They were boon companions of a dozen gamblers, suspected thieves and reputed killers of the region including such locally notorious characters as Charley Forbes, Hayes Lyons and George Ives.

Within a week they began to reveal their true character. Jack Gallagher was in the first stampede to Alder Gulch, announcing with a grin that if the gulch was as rich as reported, someone would have to be there to maintain order. Before the month was out, Hayes Lyons, Charley Forbes and Buck Stinson murdered Plummer's only honest deputy in cold blood within sight of scores of men, and Jack Gallagher, in the plot, brought them through to safety.

It was an incident of the first civil suit to be tried in the month-old camp of Virginia City, a dispute within the miners' rules which had been adopted as the only law of the camp. Dr. William L. Steele, first president of the Fairweather district, was presiding at the trial, which was hardly more than a formal arbitration. The honest deputy, Dillingham, had been designated clerk. The hearing was out-of-doors, there being no building large enough to hold the crowd. Lyons, Forbes and Stinson approached Dillingham when court recessed for a few minutes, and called him aside from the crowd. There was a moment's talk, three shots came as one, and Dillingham fell, mortally wounded.

The crowd broke into an uproar. Jack Gallagher dashed through the milling throng, skillfully substituted his own unused revolver for that of Forbes, and declared the three men under arrest. Details of what happened after that are a little vague and varied. There was no court stenographer to record the evidence, and no newspaper to set down the story. But there is no doubt that Forbes, exhibiting the unused revolver provided by Deputy Sheriff Gallagher, was acquitted, and Lyons and Deputy Sheriff Stinson convicted by a miners' jury made up of nearly all the men in the camp.

Judge Steele sentenced them to be hanged. John X. Biedler and Richard Todd were detailed to erect a gallows and dig the graves. Half-way on the short drive to the gallows, the tumbril was stopped by the screams of women of the town, prototypes of the sob-sisters of a later day, pleading for the lives of "the poor boys." "Don't hang the poor boys! Save them! Save them!" Such appeals to emotionalism, supplemented by the reading of a tear-dripping letter ascribed to Lyons and addressed to his sweet old mother, turned the tide. A new vote was taken. The voices of the women and the friends of the murderers rang through the gulch. Jack Gallagher leaped to the wagon, slashed the bonds of the killers and shouted that they were cleared. Stinson and Lyons leaped to the back of a horse, happily provided at the moment, and galloped away.

The first deliberate murder by the Plummer gang had been accomplished and the murderers freed. A sign—"Two Graves For Rent. Apply to X. Biedler"—was the temporary epitaph of justice.

D. H. Dillingham, the only honest deputy in the territory, was dead and buried. His error had been the warning of two friends, Washburne Stapleton and Jim Dodge, that Lyons, Forbes and Stinson were planning to hold them up on the trail between Bannack and Alder Gulch. Dodge had revealed the source of his warning. The news quickly reached the three desperadoes. The murder of Dillingham followed promptly. The release of his murderers through trickery and emotionalism followed almost as quickly. The Plummer gang had demonstrated both its power and its viciousness, though Henry Plummer himself was not yet suspected.

After that, robbery and murder piled upon murder and robbery. Organized crime of the most violent sort was in the saddle as it had never been in the history of the United States, and as it was not to be again until the rise of the gangsters in the prohibition era.

Within a month after the Dillingham murder, Alder Gulch was producing a fortune in gold each day. Newly organized stage companies were hauling express boxes weighted with gold over the four hundred miles of wagon trails to Salt Lake City. Individual miners, their "piles" made, were traveling back to eastern or western or southern homes with their belts and pokes stuffed full of gold. Rich pickings for bandits and bad-men. Under the efficient and still secret leadership of Sheriff Henry Plummer the road-agents organized for mutual protection and information.

Some fifty or more were in the gang. Plummer disposed them shrewdly. "Clubfoot" George Lane was assigned to open a shoe-repair shop in the first large general store of Virginia City, where he could pick up information concerning gold shipments. Ned Ray was appointed Plummer's chief deputy at Bannack. Cyrus Skinner, a saloon keeper who had been driven out of San Francisco and associated with Plummer in Idaho and elsewhere, was also a spy. His saloon was headquarters for the gang.

Billy Bunton was given the job of tending the stage station on Rattlesnake Creek, between Bannack and Alder Gulch, a position in which he could be invaluable to the gang. George Ives, "Whisky Bill" Graves, Boone Helm, "Old Tex" Crowell, Steve Marshland, "Dutch John" Wagner and a score of others were the gun-wielding highwaymen of the organization. And an organization it was, complete with a password—"I am innocent"—a badge of identification in the sailor's knot of neckerchiefs, and secret marks upon stages to denote the amount of treasure carried therein.

No sum was too small, no crime too vicious or depraved to come within the scope of the gang's activities. T. J. Dimsdale, a frontier school-teacher who wrote the history of the Montana Vigilantes while their exploits were still so fresh in the public mind that he evidently feared to give the names of most of them, has left this picture:

"Wounded men lay almost unnoticed about the city and a night or a day without shooting, knifing, or fighting would have been recognized as a small and welcome installment of the millennium. Men dared not go from Virginia to Nevada or Summit after dark. A few out of the hundreds of incidents must suffice. A Dutchman, known as Dutch Fred, was met by one of the band who ordered him to throw up his hands. Finding he had $5 in Treasury notes with him the robber told him he would take them at par, and added with a volley of curses:

"If ever you come this way with only $5 again I'll shoot you. Damn you, I'll shoot you anyhow,' and raising his pistol he shot him in the arm.

"Another man was robbed of two or three dollars about two or three miles below Nevada and was told that if he ever came with as little money again they would kill him."

So much in illustration of the smaller crimes, the terrorizing of the thousands of honest hard-working miners and business men of Alder Gulch by organized and immune criminals working under the direction of the sheriff of the district. One illustration of a greater crime, unsurpassed in brutality by even such a desperado as the historic Joaquin Murietta, will suffice for this record.

One Lloyd Magruder had brought a wagon-load of provisions from Lewiston, Idaho, to Alder Gulch in the early rush and had sold it at great profit, at retail. He was planning to return home with some twenty-seven thousand dollars' worth of gold dust and two thousand dollars in currency. He knew the dangers of the journey and planned a caravan amply able to protect itself. What he did not know was that three of the men whom he planned to take with him were members of the Plummer gang. They were "Doc" Howard, Chris Lowry and James Romaine. A fourth, Bill Page, was being broken in to the gang's needs. All four of these men had worked for Magruder while he sold out his stock of goods, and he had seen no reason to doubt their reliability. Plenty of men in Virginia City could have told him of the true character of his employees, but with the murder of Dillingham as an informer still fresh in their minds the wise men were keeping their counsel.

Besides there were to be four other men in the party, William Phillips, Charles Allen, and Robert and Horace Chalmers. Nine men, heavily armed, adequately equipped with good saddle horses, pack mules and reserve animals, should be able to make even the long trip from Virginia City through the western Rockies and over the Bitter Root Mountains in safety.

It was mid-October, and Magruder was anxious to get through the high passes of the Bitter Roots before snow fell. The party pushed on swiftly for six days before the time seemed ripe for action. "Doc" Howard, as leader of the bandits, had decided in the meantime that it would be folly to divide the loot with Plummer. Why not take it all, and move on to the coast to squander it in the brothels and gambling houses of San Francisco?

Camped in a high mountain valley, far enough from Alder Gulch to be safe from pursuit, near enough to the Pacific to make escape seem easy, "Doc" Howard fixed the night for action. Allen and the two Chalmers boys had rolled in their blankets outside the circle of firelight. Magruder, Phillips and the three conspirators were smoking beside the blaze. Page was in the shadows.

At a signal when Magruder bent to take an ember from the fire to light his pipe, Lowry split his head with an ax. When Phillips sprang to his feet in shocked astonishment, Romaine stabbed him to death with a hunting knife. Howard seized the ax, leaped over the twitching body of Magruder and split the heads of Robert and Horace Chalmers as they lay asleep. Aroused by the commotion, Allen sat up in his blankets, only to fall back dead as a charge of buckshot from Magruder's own gun, fired by Howard, blew off the top of his skull. Five men had been murdered in fifty seconds.

The three murderers congratulated one another and the timid accomplice, Page, who had had no hand in the actual killings but expected to share in the loot. This aggregated nearly forty thousand dollars when the gold dust of the other four victims was added to that of Magruder.

Then they went to work to conceal traces of the crime. The start of the first snow of the season promised to help them, covering their tracks as they carried the bodies over a near-by ridge and hurled them into a deep canyon. That done, they burned the extra equipment, scattered the ashes, and drove the surplus horses into a canyon remote from the trail and shot them.

"Doc" Howard, with a fondness for fast horses, appropriated the big sorrel of Magruder for himself. That was an error. Another error was the attempt to get through the high passes of the Bitter Roots in the midst of a continuing snowstorm. The difficulties of the journey forced them at last into the town of Lewiston to replenish their food supplies. They had hoped to avoid Lewiston, where they were known to have been employed by Magruder. But hunger and hardship drove them to it. Leaving their horses with a rancher, they sneaked into the town after dark and went to the stage station to engage passage out to the coast. Hill Beachy, deputy marshal of the town, and proprietor of the only hotel, noticed their obvious efforts to avoid recognition in the stage office, and became suspicious. Following them to the hotel, he recognized Howard, Romaine and Lowry as three tinhorn gamblers who had lived in the town a year before. He happened to know that they had worked for Magruder in Alder Gulch. He also knew that Magruder had written to his wife that he was coming home soon with the twenty-seven-thousand-dollar profits of his business venture.

Beachy immediately suspected murder. There already had been enough murders in the Alder Gulch and Bannack region to justify such suspicion. But he had no evidence on which to act, and the four men departed on the stage next morning. But Beachy was not satisfied. He inquired around the neighborhood until he found Magruder's big sorrel horse at the ranch where Howard had left it. A day or two later a traveler arriving from Alder Gulch brought the news that Magruder had left that camp, headed west, two or three weeks earlier, accompanied by Howard, Lowry, Romaine, Page and four miners, heavily laden with gold.

Beachy's suspicions were confirmed. He obtained the necessary authority and set out to capture the killers. The pursuit took him to Walla Walla, the Dalles, Portland, and at last by arduous overland journey to San Francisco. There he found the fugitives, arrested them, obtained extradition papers, and took them back to Lewiston. A confession was frightened out of the cowering Page. With that to support the circumstantial evidence, Lowry, Howard and Romaine were convicted promptly, and as promptly hanged. Page, as a state's witness, was released, only to be killed a year later in a drunken brawl. Such was life, and death, in the Northwest Territory.

The Magruder party was but one of many done to death by the road-agents of the Plummer gang. Time and again men left the Alder Gulch diggings, laden with gold, on their way to wives and families in the East, the West, the South, only to be found within a week, a month, a year, dead in some hidden canyon or shallow grave. The hard-won gold from the diggings came back to finance the orgies of the bandits, to fill the safes of the saloons, to buy silks and jewels for the painted women of the brothels and dance-halls, to corrupt what feeble authority there was.

One recognized authority asserts that a conservative count made by the Vigilantes attributed one hundred and two murders to the outlaws in a little more than three months. Another authority asserts that one hundred and ten bodies of the victims of the road-agents were found before the winter of 1863 passed over Alder Gulch. Sheriff Plummer's gang was murdering honest men at the rate of one a day, and still Henry Plummer was only beginning to be suspected.

That could not go on indefinitely even in so rich and wild a community as Alder Gulch, hundreds of miles by tortuous wagon trails from the nearest railroad. Men who had been too busy with their private affairs, mining, freighting, store-keeping and so forth, to bother about the deadly brawls in the saloons or the robbery and murder of an unimportant man leaving camp with the takings of his claim, were brought to sudden consciousness of their own danger when the news of the massacre of the Magruder party filtered back through the snow-packed mountain passes to Alder Gulch.

When the frozen body of one Nicholas Thibalt, a popular youth who had been employed in a Virginia City store, was brought into camp one day with a bullet hole in the forehead, and the marks of a rope around his neck as if he had been dragged into the brush where the body was found, public opinion crystallized suddenly. The man who had found the body offered a clue. That night an unofficial posse under the leadership of James "Cap" Williams rode out of Alder Gulch. In two days they were back with three prisoners: George Ives, accused of the murder, "Long John" Franck, eager to save his own neck by testifying against Ives, and "Old Tex" Crowell, another important witness.

Alder Gulch was alive with a new sort of interest. A murder more or less had been an incident of no great importance, but a prospective hanging of one of the badmen of the mountains, a man generally known as a vicious desperado who had hitherto escaped because of the fear in which he and his gang were held, was something to stir the populace. The gangsters immediately put their organization to work to save one of their leading men. They dispatched George Lane, their Virginia City spy, to Bannack to summon Sheriff Plummer to seize the prisoners from a lynch court and guide them to safety. As an added precaution, in which they revealed the shrewdness of the gangsters of a more modern day, they retained the four best lawyers of the district for the defense.

In that they were especially wise. Plummer, not yet definitely branded as the bandit leader, was more shrewd than his henchmen. To avoid revealing his true character either to his gang or to the honest citizens who still had faith in him, he circulated a story in Bannack that Alder Gulch had gone completely amuck. A mob was on its way to sack Bannack, murder him and his deputies, Ned Ray and Buck Stinson, and incidentally cut the throats of the highly respected Judge Sidney Edgerton and three other leading citizens. To prevent such a massacre naturally he had to stay and direct the posse which he raised to guard the town.

So Alder Gulch was relieved of interference by the cautious Plummer while it conducted its first murder trial since the travesty which had freed Stinson, Forbes and Lyons after the murder of Dillingham in the first month of the camp's history. This trial, like the other, was held in the open air, although the date was December nineteenth and the weather bitterly cold.

Two wagons were drawn up in the street to serve as judge's bench, counsel's table, and witness stand. Benches from a near-by dance-hall were placed to accommodate a jury. A huge bonfire blazed for the intermittent comfort of frosty hands and feet.

Don L. Byam, miners' judge, whose chief duty up to that time had been the settlement of disputes concerning claims, presided. Wilbur Fisk Sanders, a bearded youth of twenty-nine who had been admitted to the bar in New York, was appointed prosecutor. It was not a job to be envied, with dozens of the defendants' associates, armed with revolvers, shotguns and rifles, circulating through the crowd which jammed the frozen street for a block. They swaggered, threatened and boasted that no such court, without authority and without power, could prosecute their friends and live to tell the tale. But Wilbur F. Sanders was a man of nerve as well as trained intelligence. He lived not only to prosecute George Ives and to tell the tale, but to prosecute a score of murderers in trials equally impromptu, and to be elected the first United States Senator from Montana.

A jury of twenty-four men, their courage fortified by Sanders' unwavering prosecution under the hidden threat of half a hundred guns, declared George Ives guilty of the murder of Thibalt. Sanders promptly moved that the verdict be recorded as "the verdict of the miners' meeting here assembled," and that Ives "be taken to a suitable place and there hanged by the neck until he is dead." A roar of "Ayes" from a thousand men drowned the "Noes" of the few who sought to retain their immunity to rob and murder. George Ives was hanged from a projecting log of a building in process of construction near by. The first step had been taken to bring the safety of civilization into Alder Gulch.

Within a week the Vigilantes were organized to carry on the good work. Paris F. Pfouts was president, James Williams executive officer, Wilbur F. Sanders prosecutor and John S. Lott treasurer.

Testimony elicited at the trial of Ives pointed unerringly to several other men who were widely known for their reckless spending of gold dust, without evident source of supply. Encouraged by the temporary triumph of order without law, witnesses named "Dutch John" Wagner and Steve Marshland as the men who had held up the Moody pack train, carrying seventy-five thousand dollars, three weeks earlier. "Cap" Williams set out secretly with a Vigilante posse to seize them and Alec[k] Carter, who had been implicated in the Thibalt murder.

Near the outlaws' rendezvous, a log ranchhouse, still pointed out to tourists as "Robbers' Roost," the posse encountered a man of the district known as "Red" Yeager. Yeager was not suspected of association with the bandits. When questioned by "Cap" Williams he answered cheerfully that the men wanted were carousing in Deer Lodge with the notorious "Whisky Bill" Graves and Billy Bunton. When the posse arrived at the rendezvous they found that their quarry had been warned by a note from one George Brown, brought by Yeager ahead of the posse with such speed that he ruined two horses on the journey.

The Vigilantes promptly turned back and took both Brown and Yeager into custody. A brief hearing established their guilt as accessories to the flight of the other desperadoes and they were sentenced to hang. The honest citizens of Alder Gulch had been slow to awaken, but they were wide awake at last.

Facing the noose, "Red" Yeager confessed his complicity in a dozen robberies, and named Sheriff Henry Plummer as the chief of the bandits. That was the first time the sheriff had been definitely accused, although he had been under suspicion for several weeks. Going the whole route, once he had started, Yeager dictated the names and duties of more than a score of men in the Plummer gang, and added details of a score of robberies and several murders.

That information given to the posse, Yeager went calmly to his death at the end of a rope over the limb of a convenient tree, wishing the posse luck in the punishment of his evil associates. Brown, whining and pleading for mercy, died beside him. For five days the frozen bodies swung in the winter wind.

As fast as they could ride over the frozen trails and snowdrifts, the Vigilantes rounded up "Dutch John" Wagner, Ned Ray, Buck Stinson, and finally Sheriff Henry Plummer himself. Within a few hours of each arrest, all the evidence contained in the Yeager confession had been substantiated against each man. Dozens of honest citizens who had feared to tell what they knew while the sheriff and his road-agents were in control of the situation came forward to give information to the Vigilantes. Within a day all of Plummer's vicious record was disclosed. The verdict for Plummer, Stinson and Ray was death. One after another they dropped to the ends of ropes on a scaffold erected by Plummer himself for execution of a murderer legally convicted some time earlier.

Joe Pizanthia, another of the desperadoes, resisted arrest and shot a member of the posse. Within an hour he was hanging, his body riddled by bullets. "Dutch John" was next, hanged from the rafters of an unfinished house where the bodies of Plummer and Stinson still lay, awaiting burial.

Returning to Alder Gulch, the posse rounded up Hayes Lyons, Jack Gallagher, Boone Helm, Frank Parrish and George Lane. Lyons, it will be recalled, had escaped the noose in Alder Gulch only six months before, through the connivance of Gallagher and the sobbing sentimentalism of a few women, thereby fixing in the minds of the desperadoes a sense of immunity which was the real foundation of the gang. But that immunity had reached its end. The five men were hanged in a row from the rafter of an unfinished frame building in Virginia City.

Such convincing evidence that their rule of the diggings was at an end scattered the remaining bandits quickly. Why Plummer and his deputies had not fled from Bannack or Hayes Lyons and his associates from Alder Gulch at the first word of the confession and hanging of "Red" Yeager is a mystery. It may have been that the completeness of their sway through six months of unpunished robbery, murder and debauch, had turned their heads. But now, with their sheriff chief and a dozen of their associates thrown to the final punishment, those remaining began to take to their heels.

Some of them did not go far enough along the frozen snow-blocked trails. "Cap" Williams, leader of the posses, was as persistent as Joubert. Wilbur Fisk Sanders, prosecutor, was as implacable as Fate. Bill Bunton, Cyrus Skinner, Aleck Carter, Bob Zachary, Johnny Cooper, George Shears, "Whisky Bill" Graves, Bill Hunter, were trailed to their hiding-places, confronted with adequate evidence of their guilt, and strung from gate posts, trees, ladders or whatever was most convenient.

In the single month from January 3 to February 3, 1864, twenty desperadoes were summarily hanged. At least one had been hanged before; at least one later. The far-famed San Francisco Vigilantes could boast of no such record as that.

When the half-completed list of executions had been read to Alec[k] Carter, standing on the gallows with a noose around his neck, he had said; "All right! Not an innocent man hanged yet!"

Montana had cleaned up. More than a score of men, of a gang numbering perhaps fifty, guilty of the murder of more than one hundred, had paid with their lives. The others, in terror, had vanished from the mountains.

A civilization which had temporarily permitted the theft of some thousands of dollars' worth of gold dust from the millions being mined in Alder Gulch, had answered violent death with violent death. It had cleared the way for advancement to a more effete civilization which would justify fraud running into countless millions, a score of suicides and more violent deaths to the accompaniment of production of more than two billion dollars. The way was cleared to Butte, and Wall Street.

CHAPTER III

PREPARING TO OWN AMERICA

WILLIAM ANDREWS CLARK, a driving, grasping, ambitious youth of truly astonishing energy, and business ability near to genius, was preparing to play a leading rôle in one of the greatest economic, political and social dramas in the history of the United States. From the level of a laborer at two dollars and fifty cents a day in one of the early mines of Colorado, Clark was on the way to becoming, according to the late Senator Robert M. LaFollette, "one of a hundred men who own America."

Marcus Daly, equally shrewd, equally energetic, but otherwise the antithesis of Clark in every phase of character, was learning the mining business, also with pick and shovel. His first experience was under John Mackay and James Graham Fair in the historic Comstock Lode of Nevada. He was to win many millions from the vast copper deposits of Butte which he was first to vision in all their immensity. He was to see these millions watered in Wall Street and doubled through promotion of the Amalgamated Copper Company. He was to stop at that point in his rôle of empire-builder only because death intervened, but the copper empire which he started was to control Montana and dictate on occasion to the United States.

F. Augustus Heinze, the most dramatic figure in the war for millions and for power, was a child when Clark and Daly were men, but one eventually to wring their souls and cut into their profits to a cry of "Napoleon! Napoleon!" from ten thousand fellow townsmen of Butte. Heinze was growing up in Brooklyn to be cursed by Standard Oil magnates as the most vicious enemy ever to oppose their theory of advancing civilization. The stage was being set in the 'seventies for a tremendous drama.

Clark was of the second generation of Irish immigrants who had come to the United States in Revolutionary times. He was born on a farm near Connellsville, Pennsylvania, January 8, 1839. At that time there were only twenty-five states in the Union. Mexico still owned what is now New Mexico, Arizona, California, Nevada and Utah. Texas was an independent republic. England still claimed the territory of Oregon, Washington and part of Idaho. Virtually all the vast area drained by the Missouri River was ungoverned Indian country known only to a few white trappers, traders and pioneer explorers. Half of the present area of the United States was wilderness. It was a land of opportunity. William Andrews Clark grew up to seize that opportunity.

Marcus Daly, born in County Cavan, Ireland, December 5, 1841, had come to America as a child, sold newspapers in New York, and worked his way to California in time to join the rush to the Comstock Lode in 1860. There he was preparing to become an able antagonist of the able Clark in the struggle to own and rule "the richest hill on earth."

Before these battling Irishmen had started their epic struggle, a lad of German Lutheran parentage was learning his letters in the schools of Brooklyn. He was to make the battle of Butte a battle royal with uncounted millions as the prize, and more scandal, libel, corruption and vituperation involved than any financial and political war had ever inspired upon this continent. That man was F. Augustus Heinze, born in Brooklyn, December 5, 1869, twenty-eight years to a day younger than Marcus Daly, nearly thirty-one years younger than William Andrews Clark.

It is doubtful that ever in the development of civilization in America have three more diversified characters met in more strenuous struggle. With a financial prize beyond computation, no pick-and-shovel man could be permitted by an advancing civilization to walk away with the profits. While the unremitting labor, the cheerfully endured hardships of Clark and Daly were opening the way to the goal, the advantages of civilization were training Heinze in the scientific theories of mining engineering and metallurgy. Simultaneously John D. Rockefeller's shrewd brain and tractable conscience had brought the Standard Oil Company into the position of the richest and most powerful monopoly in the United States. Incidentally it had developed Henry H. Rogers and William Rockefeller into able lieutenants.

Clark, Daly and Heinze each started, so to speak, at scratch, though at different dates. Clark entered the Montana territory in 1863, within a year of the discovery of the rich Grasshopper Creek placers, and a few weeks after the discovery of the one-hundred-million-dollar diggings of Alder Gulch. Daly arrived in 1876, when Butte was nothing more than an insignificant silver camp. Heinze's entry was delayed until 1889, when the fame of "the richest hill on earth" was being heralded around the world.

But each of those men entered the region with little more than his hands and brains to make the battle. Both hands and brains they used in person. The Standard Oil magnates, in contrast, armed with the power of many millions of dollars and attendant vast political power, sat in New York offices, as aloof and impersonal as a general staff, and directed their hirelings to do the dirty work. What came out of the fluctuating alignment, including the national financial panic of 1907 and more than two billion dollars of material wealth is history, and the subject matter of this story.

Clark was the pioneer, advancing along lines set by himself with the development of the great West and the advancement of civilization, or its evidences, in America. Educated in the rural schools of Pennsylvania as far as such education went up to in 1856, he moved with his parents to Iowa. There he gained what additional education was available in Birmingham and Mount Pleasant. Probably it was not much, but enough to make him a country school-teacher in Missouri, where he came into touch with many men setting out with high hopes for the mines of California and Nevada. When the mineral riches of Colorado were discovered he was eager to go, and in 1862 he moved to Colorado.

Neither work nor hardship held any terrors for Clark. When news of rich gold discoveries in what was then known as eastern Idaho, now western Montana, filtered into Central City, Colorado, where Clark was mining, he decided to go at once. With a group of his pick-and-shovel associates he journeyed to Denver, purchased a wagon and two yoke of oxen, food and mining supplies, and set forth to the new diggings.

That was the year in which Lincoln freed the slaves, the year of the Vicksburg campaign, the year in which General Robert E. Lee crushed the Union Army at Chancellorsville, the year of Gettysburg. If Clark had been of different character he might have found adventure as many thousands of other young men of his age were finding it—in fighting for his country. But Clark preferred his adventure with a promise of cash profit. In his own words, describing that period of his life, "all of us were imbued with the same ambition, to endeavor to better our condition in the world if possible." It was an ambition which remained as dominant with Clark when he had accumulated fifty million dollars as when he had invested his entire capital in the project to invade the gold diggings near the continental divide.

In 1863 the inspiration of gold in the development of the West was fresh and dominant. Only the fact that he had been ten years old instead of twenty had kept William A. Clark from participating in the California gold rush of 1849. Only the fact that his savings as a country school-teacher were pitifully small had stopped him in Colorado instead of carrying him on to the more sensational Comstock discoveries of western Nevada.

Yet he was true to type, even to the ox team and equipment with which he and his three companions set out to "better their condition" in the gold placers of Grasshopper Creek. Evidence of the nature of the trackless wilderness through which they journeyed is revealed in the fact that sixty days were required for the trip. Passing through hostile Indian territory, they joined a larger wagon-train for mutual protection, and trudged wearily into Bannack.

They had hardly unyoked the oxen when news of the discovery of rich new diggings within a day's walk stirred the camp. Wearied as he was, Clark struggled on without a moment's hesitation to the new discovery, incidentally paying his way by hauling a trader's supply of liquor and tobacco. Two of his associates had preferred to go on to the more fabulously rich diggings of Alder Gulch. Clark and his remaining Colorado associate acquired a claim in Jeff Davis Gulch, near Horse Prairie Creek. That done, Clark returned to Bannack to buy lumber for sluice boxes. His reminiscences as told before the Society of Montana Pioneers sixty years later cast an illuminating light upon this early period of his career.

"We were not in very flush financial condition at that time. Upon my arrival at Bannack I found five letters from home that anticipated me and had been carried from Salt Lake City by a private express that had been established between that place and Bannack. The price of transportation of a letter at that time was $1.00 each, and I had just $5.00 value in Bob Tail dust—a term applied in Colorado to gold amalgam, the product of the mills of that country, and at the time it was the sole currency in circulation in Colorado. I had, besides, a fractional greenback currency, of the denomination of fifty cents.

"I gladly dispensed with the $5.00 for the letters. Therefore I was obliged to get credit for the lumber and some few other articles which we needed, and this I readily obtained. During our prospecting trip I had found a very fine pair of elk antlers, which I brought into Bannack, and for which Cy Skinner, who kept a saloon and who was afterwards hung by the Vigilantes near Hell Gate, offered to give me $10.00, and this I readily accepted.

"Returning to the gulch, I found that my companion had commenced operations, and it was not long until we installed our equipment and began our first experience in placer mining. We first got some logs and built a cabin of about 15 feet by 20 feet dimensions, which was covered in the conventional style with a roof of split poles covered with dirt, and which we found very satisfactory in absolutely dry weather. We took the hind wheels off the wagon, which we converted into a cart for the purpose of hauling dirt to Colorado Creek, where we constructed three sluice boxes of the lumber which I had brought from Bannack. . . .

"At the close of the season we closed mining operations and put everything in order to leave the camp. We had paid up all our obligations and had several thousand dollars each in gold dust left. . . ."

That several thousand dollars, gained by manual labor upon his own mining claim was the beginning of Clark's colossal fortune, and the first great stimulus to his overpowering ambition. Never again in the succeeding sixty-two years of his life was he to be in anything approaching the financial condition in which he had found himself when he sold the set of elk's antlers to Cy Skinner, the saloon-keeping highwayman. But neither did his growing wealth ever check his acquisitive energy to undertake any labor, danger or deal which promised a profit.

Most of the miners of his day with several thousand dollars in their pockets, facing a frozen winter in such a camp as Bannack, amply provided with saloons, food, gambling and women, would have settled down to unrestricted enjoyment of the coarse entertainment available. Not William Andrews Clark.

He promptly found a job of cutting and hauling firewood from the near-by hills to Bill Goodrich's small hotel for two dollars a day and his board. But though physically able to withstand hardships, Clark drew the line at personal peril, when he could. When the third day of his wood-cutting job trapped him in a blinding blizzard from which he narrowly escaped with his life, he quit the job. Then he induced a few friends to invest with him in a team and wagon each, and go to Salt Lake City "and take a look at the Mormons, concerning whom I had heard many interesting stories, and to buy something appropriate to the mining camp, which we might bring back, and thereby make expenses and possibly something more." That was William Andrews Clark, then and forever.

The journey of four hundred miles over primitive and sometimes dangerous roads was made in twelve days. Clark found the Mormons and their capital interesting. "There were probably not to exceed ten thousand people in the city, but it was laid out on broad lines, as you who have visited it have undoubtedly observed. There was only one public hotel, the Salt Lake House, and it was a very poor one, consisting of a frame building with limited accommodations. They had a very commodious theater, however, and some of the local actors had remarkably good talent. They had already begun the foundations of the Temple, and great blocks of stone had been quarried and lay around it. . . . The favorite beverage of the inhabitants, which they called 'Valley Tan,' was a colorless species of whiskey, and the taste was abominable. We observed that many of the Mormon girls were very pretty."

A few days' acquaintance with Salt Lake City, Valley Tan and the pretty Mormon girls, however, satisfied the party. Each man loaded his wagon with commodities which he believed might be in demand at Bannack or Alder Gulch, and started back through the winter dangers of the Rocky Mountains. It was an arduous journey. "Crossing the Divide," Clark reported many years later, "I witnessed what I had never dreamed of before—several cattle in the moving train freeze to death in the yoke and go right down upon the ground."

In spite of such difficulties they arrived in Bannack in time to see "Dutch John" Wagner, one of the first men taken by the Vigilantes, hanging, frozen, from a rafter in an unfinished building of the camp.

"We proceeded in due course to dispose of our merchandise at very good profits. I had taken the risk of shipping quite a lot of eggs, well knowing they would freeze, yet they were admirably adapted for the making of 'Tom and Jerry,' which was a favorite beverage in Bannack, and I disposed of them at a price of three dollars per dozen."

With the break-up of winter Clark returned to his mining claim, operated it profitably for a few months, sold it, then headed back toward Bannack. Ordinarily it would have been a day's journey, but illness seized Clark on the way. According to his own reminiscences he was taken in and nursed by a French couple who owned a ranch on the road. Their care and kindness, and the skill of a physician summoned from Bannack, he says, saved his life.

There is still current among the old-timers in Butte and Helena another story of that illness which reflects more creditably upon Clark's character than his own account. According to the more popular story, the fever-stricken man was taken in by a rancher or prospector named Brown, later to be known as "Crazy Brown." Brown nursed Clark until he began to despair of the patient's life, and then carried him to a doctor who brought him safely through the crisis of pneumonia. In after years, according to the popular tale, when Clark was operating banks in Deer Lodge and Butte, he left a standing order with his paying tellers that any draft for a reasonable sum presented by his benefactor should be honored, and charged to his account. According to the same legend, however, Brown or the Frenchman, or whoever it was, never availed himself of the opportunity for easy money as a reward for Good Samaritanism.

In any event, Clark recovered from his illness. His first year in the new mining camps of the Montana wilderness had passed successfully, and his pick-and-shovel days were at an end. Another merchandising trip to Salt Lake City followed. An ox-train of goods was dispatched from Salt Lake to Alder Gulch, and Clark went back by stage to meet it.

"At this time (1864) the stage coach in operation between Utah and Montana was a very primitive and unsatisfactory affair, but a few years afterward it was much improved and afforded a more speedy and comfortable passage. The trip from Utah to Helena and other Montana points required only about six days' and nights' constant travel, usually with six horses or mules to each coach. Changes of the animals were made about every twenty-five to forty miles, with fairly good eating stations along the line. There was no regularity as to sleeping. Short naps were the rule, but the aggregate in twenty-four hours was sufficient for most of the passengers. The stages used were of the Ben Holliday type, accommodating nine passengers inside and one on the box with the driver."

Waiting for the arrival of his heavy freight train in Virginia City, blocked by deep snowdrifts, soon palled on a man of Clark's nervous energy. He journeyed to Bannack to dispose of the goods which had reached there, and incidentally to attend the meetings of the first legislature of the Territory of Montana. "The meetings of the legislature were very interesting," he says, "and I spent several days there very pleasantly. The first legislative assembly was held in a log cabin of modest dimensions, but as the entire legislative assembly comprised only twenty members the space seemed sufficient, although the room for the lobbyists was somewhat limited, and lobbyists were in evidence in those days as well as afterward."

In that first session of Montana's territorial legislature was Clark's introduction to practical politics, arousing an ambition which was to carry him along a devious and scandalous route to the United States Senate. Incidentally, it introduced him to the newly founded mining camp of Helena, which he was to make the capital of the state, through a political battle marked by expenditure of more than a million dollars and a vicious warfare with Marcus Daly which lasted until the day of the latter's death.

John Cowan had just discovered gold in Last Chance Gulch, the line of Helena's present Main Street. A man of Clark's character could overlook no such opportunity. With two old miner acquaintances he set out by wagon, camping in the snow en route. He found a few houses and cabins being constructed in the new camp, and "an air of activity which augured well for its future."

Thereafter, Clark's interest in Helena never waned. But the more pressing business of his ox-train of merchandise stalled in the mountain snows on the way to Alder Gulch called him back to Virginia City. The price of flour was climbing to one dollar a pound in the snowbound camps of Alder Gulch. All through the mountains many persons were living on "beef straight," or upon such wild game as they could kill.

It was a winter of hardship and deadly monotony, a spring in which virtually the only news to reach the mining camps was the report of the assassination of President Lincoln. Clark was still waiting anxiously for the arrival of his merchandise when the owner of the ox-train arrived in Virginia City on snowshoes and announced a plan to make deliveries by substituting mules for the oxen and sleds for the wagons to surmount the snow. It seemed an opportunity for tremendous profit, as the thousands of people in Alder Gulch were eager to pay five and ten times as much for the food and clothing as it had cost in Salt Lake City. But here again Clark revealed his shrewdness.

"The project seemed feasible," he said later, "and some, if not all of those interested made terms with him (the ox-train owner) on the basis proposed. I had had an experience with mules the winter before on the same route, and found that when Mr. Mule got down in the snow he would not get up until he was dug out. I was also anxious to make plans for the summer, and made a proposal of sale of my merchandise to him at prices very much below those prevailing, which seemed to him atttactive and the transaction was closed.

"I received part payment of the amount agreed upon and the balance was secured by his banker. He bought sixteen fine large mules and started at once to put his plan into effect. It was necessary to haul feed for the animals which involved considerable expense. He encountered many difficulties, and unexpectedly winter broke up, the bull teams began to come over sooner than expected, and prices rapidly declined as a natural consequence."

What the freighter's losses amounted to is not disclosed. But Clark's profits were sufficient to justify a stage journey to Fort

Benton on the Missouri River, down the river for the first visit with his family in Iowa in four years, and a tour of Boston, New York, Philadelphia, Washington and points south. Even on a pleasure trip of such great extent for that early day, Clark did not step out of character.

"In order to help pay expenses I had, upon the advice of a friend in St. Louis, purchased a few barrels of pork and some other staple goods in the South, and consigned them to myself in Memphis. I spent a few days very pleasantly in that prominent southern city, but found a dull market upon arrival of my goods, and I therefore reshipped them to Canton, which was a city of considerable importance in Mississippi. I followed by rail and spent a few very pleasant days at Canton, and was very quickly able to sell my little stock of goods to one merchant."

The grand tour ended with another brief visit to his family home in Iowa, and return to Montana by way of Denver and Salt Lake City. Montana was thriving though winter had closed down when Clark arrived again in Helena. Last Chance Gulch and its neighboring Grizzly Gulch had been producing richly from the placers. Gold was coming in vast amounts from Alder Gulch. The Grasshopper Creek diggings were doing well. Business opportunities revealed themselves to Clark on every side.

The first and most promising seemed to be a contract for carrying the mail between Missoula, Montana, and Walla Walla, Washington. Clark's travels had made him familiar with the route. His organizing ability and the ready capital available to buy horses and establish stations on the mountain trails quickly completed the arrangements. His brother, Joseph K. Clark, was put in charge of the eastern end, and Warren Whitcher of the western. For the first time Clark began to take profits from the labor of employees. It opened a new field to his ambition and energy. He overlooked no bets.

"In addition to acting as agent for several merchants in Helena, I also made loans at a rate of interest ranging at that time at about two per cent a month. There were many opportunities for advantageous purchases of gold dust, of which I availed myself, and there was no lack of opportunities for those who were on the alert for making money. For instance, Dan Floweree had about 160 head of cattle which he had used in freighting goods from Fort Benton, and also from Salt Lake. There was a lull in the business, and freighting became unprofitable. Finding that the cattle were for sale, I looked them over and purchased them for a lump sum, and had them driven to Nevada Creek Valley, where I knew the grass was high and luxuriant. After a few weeks they were fit for the market, and I sold them out in lots of eight or ten at a time to the butchers in the surrounding mining camps."

And so it went with William Andrews Clark. Business, for the time being, promised more profits than mining. Buying gold from the miners was a more dependable source of income than digging it. Darius Ogden Mills had proved that most effectively in California. Within the next few years Clark's merchandising activities broadened rapidly and profitably. By 1869 he was established in Deer Lodge in a wholesale merchandising and banking business under the title of Donnell, Clark and Larabie. He had married Kate L. Stauffer, a childhood sweetheart of his schooldays in Pennsylvania, and was finding capital available to take a renewed and greater interest in mining.

The way was cleared by experience and business genius, and the hardening of youthful character, for a sensational career.

CHAPTER IV

BUTTE GETS UNDER WAY

IN THE meantime the sensationally rich gold diggings of Alder Gulch and Last Chance Gulch had lured thousands of prospectors and miners from the looted placers of California, from the temporarily diminishing returns of the Comstock Lode, and from the undeveloped Colorado district. Hundreds of these prospectors, too late to stake a claim on the rich gravel of Alder and Last Chance Gulches, scattered out through the mountains to seek unpreempted riches.

So, in May of 1864, Budd Parker, P. Allison, and Joe and Jim Esler washed some gravel in Silver Bow Creek, a few miles southwest of what later became the city of Butte. They found it promising. The country was swarming with prospectors, and close upon the heels of the first group came G. O. Humphrey, Denis Leary and others. The camp of Silver Bow was founded, and prospered briefly. But it was not in the same class with the great diggings of Alder Gulch.

Within a few weeks G. O. Humphrey and William Allison disposed of their claims and set out toward the mountain which dominated the country, only a few miles away toward the northeast. On that slope an abandoned prospect hole a few feet deep, containing the remains of some elk antlers which appeared to have been used for digging, stirred their interest. They deepened the hole with their shovels, carried some of the dirt to the creek below, and washed out a few grains of gold. That was the beginning of Butte—a scratch at the foot of "the richest hill on earth."

Denis Leary, H. H. Porter, Charles Murphy and a few others soon joined the discoverers. A mining district was organized with Allison as president and Humphrey as recorder.

Humphrey, Leary, Porter, Tom O'Connor and George Newkirk built a cabin when winter threatened, and the town of Butte was founded, though unsurveyed. But the placers were meager, the gold extremely hard to recover, and the camp, after attaining a population of five hundred in two years, began to fail. When, in 1866, the first townsite was surveyed, one disgruntled resident voiced the opinion of the majority. "Yes, they've laid out Butte at last," he said, "and if it has any friends now that it is laid out, they'd better come and bury it."

The group which had built the first house, on what later became Quartz Street, broke up. They considered themselves lucky to sell the property, including a vacant lot adjacent, to a man named Girton. The placers were exhausted after a total production of about one million five hundred thousand dollars. The town was reduced to a few dozen men, striving valiantly to do something with the quartz which their digging had revealed.

John Noyes and David Upton had built a water ditch twelve miles long at a cost of twenty thousand dollars to bring water to the claims. Denis Leary and T. C. Porter built a smelter which failed to smelt. Joseph Ramsdell and William Parks built another which was no better. Butte was dying, almost before it had been christened. But the type of men who were its founders refused to recognize the illness as fatal.

The Ramsdell brothers, the Porter brothers, Denis Leary and George Newkirk located what was to become the Parrott Lode, and did the assessment work, sinking a shaft into quartz. W. L. Farlin located the Travona Mine, and freighted its ore by ox-team over two hundred miles of primitive roads to Fort Benton, and thence by river and rail to a smelter in New Jersey. There was no profit in that. He tried four hundred miles by wagon to Corinne, Utah. There he was rewarded with the information that there was four hundred dollars' worth of bullion, mostly silver, hidden in each ton of his ore. But the freight and smelting costs consumed all that. Still the remaining mine owners, most of whom were working with pick and shovel, struggled on, desperately determined to succeed.

These were the men, this the situation in Butte when William Andrews Clark, at the age of thirty-three, risen to the position of the leading banker, merchant and capitalist in the Territory of Montana, visited the camp in the summer of 1872. Not one man of the score who remained out of the earlier population of five hundred had the slightest doubt that his mine contained a fortune. The only problem was to get it out. All that was needed was capital to work the refractory silver ores.

Clark represented capital, and capital, with America going down the chute toward the devastating panic of 1873 was extremely hard to come by. The miners welcomed Clark with open arms and led him to the outcropping ledges which crossed the camp and the hill above. There, they insisted, was a vast fortune awaiting his grasping hand.

Clark was precisely the sort of man to listen to that sort of talk, and to examine those ledges and the shafts which had been started into them. He had made money in Montana, but far from enough money. Time was to prove that even when he had a million dollars, ten millions, fifty millions, he still did not have enough.

His banking and allied business activities had settled into a routine which gave him time and liberty. But his profits were falling off through the decrease in production of Alder Gulch and Helena gold, and the national depression following the Civil War. His firm had been buying from the miners and selling to the Philadelphia mint one million five hundred thousand dollars of Montana gold each year, but that was not enough, although the profits on that branch of his business alone averaged something like one hundred and fifty thousand dollars annually. He sought new and greater opportunities. He found them in the undeveloped mines of Butte.

But he was cautious as well as shrewd and exacting. He had occupied some of his time from the days of his first mining experience in Colorado in the study of geology and mineralogy. In fact he had carried with him on his first arduous sixty-day journey from Colorado to Montana a copy of Hitchcock's *Elements of Geology*, with a copy of *Parsons on Contracts* and *The Poems of Robert Burns* for lighter reading. He had been a student since childhood, and his contacts with mining men through the last twelve years had concentrated his studies upon mining and metallurgy. So it was no ignorant greenhorn, ripe for picking, upon whom the few working mine owners of Butte turned their sales talk.

Clark listened and looked. He examined carefully the ore exposed in the few shallow shafts, and decided that it might be worked at a profit. The price set upon their holdings by some of the hard-pressed owners was pitifully small. Clark left Butte with options on a controlling interest in four claims—the Colusa, Original, Mountain Chief and Gambetta.

At the same time he decided to go straightway to the Columbia School of Mines in New York for an intensive course of study. Incidentally he would take specimens of the Butte silver ore with him for assay and analysis, with a view to development of a practical method of working the ledges at a profit. Butte was still some years from profits, but Clark was prepared to take advantage of any developments.

W. L. Farlin, with quantities of four-hundred-dollar ore in his Travona Mine, and the unprofitable experience of freighting that ore by ox-train to far distant refineries, managed to get enough money together by 1875 to build a stamp mill and refinery on the ground for treatment of his ores. Clark advanced thirty thousand dollars on the property. Then, and not until then, was Butte really on the way to riches and fame, and that as a silver camp.

Simultaneously the Centennial mill was under way. Clark's Dexter mill was completed in 1876. Farlin, unable to meet his obligation to the Clark bank, allowed the property to go. Clark's partner, Larabie, was not interested in mines. Clark bought his rights for a drove of horses. Silver ore assaying four hundred dollars a ton, even when worked by the crude methods available in Butte returned a goodly profit. The city grew swiftly from a population of fifty to five thousand. Eastern capital and eastern newspapers began to take an interest. Still Butte was hundreds of miles from a railroad, and the newspapers described it with a free hand and imagination unrestrained. They had acquired ideas from Virginia City, on the Comstock Lode.

One eastern publication eventually asserted on the authority of what it called "reliable sources," that Butte was "simply an outpost of hell; that men lived by systematic robbery of people enticed into the camp by paid-for accounts of the fabulous riches of its mines; that murders were of such frequent occurrence that the few women and children there looked with indifference upon crime of every kind, and often joined the men in its commission. As for law and order, they exist only in the memory of those who have seen better days."

Another inspired reporter wrote: "Every business man in Butte and every miner is a walking arsenal. He carries a brace or two of pistols in his belt and a bowie knife in his right boot. Broncos are ridden into public and private houses, as it suits their drunken riders. Men, women, negroes, Chinese and Indians daily and nightly congregate in one common assemblage around the gaming tables, with which the dissolute, hilarious camp abounds."

That was the picture of Butte which was broadcast to the world. The fact that it was a false picture made it no less embarrassing to the respectable citizens of the rising town when business called them to eastern cities. When outsiders came in, stepping cautiously to avoid expected gunplay, Butte laughed philosophically at their temerity and showed them its mines. The mines and mills by that time were assuring a prosperity for many years to come.

To be sure, there were dance-halls, gambling houses, a red-light district, occasional crime, and other appurtenances of a rich and rising frontier mining camp, but there were also churches, schools, family life, and jobs for everyone who was willing to work.

The new town resting against the foot of the big hill, above the semicircular flat through which Silver Bow Creek meandered, had even a sort of bleak beauty of setting. The first cabins built in Town Gulch, which was later to become infamous as Dublin Gulch, were sheltered by pine trees. Higher on the slope was excellent pasturage for the horses, mules and oxen which provided the town's only method of transportation to a world which was just beginning to hear about it. Below, in the flat, grass grew knee high, free to any animal that might seek it.

There was even a hotel of sorts, the most pretentious building in the town, the Hotel de Mineral, with the accent distinctly upon the last syllable. It was owned by John and Simon Hauswirth, of a family still active in the life of Butte. The town was on its way. But the ores of its mines were refractory. A little gold, some copper, some zinc and a little sulphur mixed with the predominating silver made them difficult to work. Much of the value was being lost in the mills, but there was money coming in, and word of greater riches going out.

W. A. Clark, returning from his intensive study of mining engineering and metallurgy in New York, armed with a more scientific understanding of the local problems than most of the so-called practical miners of the day, was able to make considerable profits from his mines and mill. At twenty-five and thirty dollars per ton, the milling profits were astonishing. Clark secured his controlling interest in the Colusa, Original, Gambetta and Mountain Chief. He located the Frank Moulton, just west of a claim that was coming into production as the Alice Mine, near the top of the big hill. Business was good. The word of a silver town that promised to rival Virginia City on the Comstock Lode spread far and wide.

Walker Brothers of Salt Lake City, leading bankers, merchants and promoters of the Mormon metropolis, the same Walker Brothers who had purchased a wagonload of tea, tobacco and whisky from "Lucky" Baldwin in 1853, sent a practical miner to investigate the possibilities of the new camp. He was ordered to examine the Alice particularly. Owners of that mine had been buying supplies from the Walkers and paying for them with silver bullion.

The investigator was Marcus Daly, who was soon to become the most bitter and consistent enemy William Andrews Clark was ever to encounter. It is an illuminating sidelight on Clark that when, half a century later, he prepared his reminiscences for the Society of Montana Pioneers he ignored the name of Daly, referring to him only as "a representative of Walker Brothers."

Daly, with years of practical experience in the silver mines and mills of the Comstock, seemed to the Walkers to be the best man available for their investigation. So-called practical mining men in that day were looked upon with far more confidence than the geologists and engineers issuing from the few limited mining schools of the country. Daly came to Butte by stage from Salt Lake City and made his examination of the Alice, with an incidental investigation of the neighboring properties. A local newspaper of August 25, 1876, gave him passing comment:

"Mr. Marcus Daly, a western miner of experience, and a gentleman of good sense who has been looking around the quartz camps for some time, left by coach for Idaho yesterday, and we trust will see it to his interest to come back and take a hand at mining in Montana."

How greatly to his interest and to the interest of Montana it was is indicated with startling emphasis in the spreading headlines of another Butte newspaper, twenty-four years later:

A MIGHTY OAK HAS FALLEN
The Architect of Montana's Greatness Is Gone
Marcus Daly Is Dead
His Name and Works Held Sacred in Montana—Love That Was His Due in Life Now Made Manifest—Marcus Daly Was a Gift of Nature—Greater Than Napoleon—A Leader of Men—Died Amid the Monuments of His Glory—All Montana Mourns His Death

But in 1876 Marcus Daly was merely an Irish miner with a warming smile, a genial handclasp, a nose for ore, and the job of reporting to Walker Brothers as to the potential value of the Alice Mine. That report was favorable. And therein lay the inception of the famous Clark-Daly feud, a feud that was to result in the expenditure of vast fortunes by the participants, broadcast charges of bribery and corruption, and involvement of Clark in such scandal as few men survive.

There are a score of stories available as to the inception of this long and bitter battle. The one that goes back farthest is told by Ben E. Stack, in 1876 living in a cabin near the Alice Mine in Butte, in 1934 living in a cottage in Long Beach, California, and still able to give convincing evidence of the accuracy of his memory.

"Marcus Daly lived not far from me," says Stack, "and frequently came to my cabin to read the *Irish World,* and to chat. He told me the story. He took an option on the Alice from Rolla Butcher and gave a draft on the Clark and Larabie bank, in Deer Lodge. Clark had an eye on the Alice but Daly's option upset his plans and he refused to honor the draft, saying Daly had no authority. Butcher told Daly who then gave him an express order through Wells, Fargo. Butcher collected on this order, and the feud was started."

That is the earliest story of the start of the feud. Most of the others put it at a later date, and ascribe it to difficulties growing out of the first Anaconda smelter. The one of these most generally accepted is that when Daly had made his plans for building the smelter he found that Clark had filed on the water rights of Warm Springs Creek, which Daly had expected to use. Water was essential to the development of the project. Clark, according to this story, put the price of his water rights at an almost prohibitive figure and, having the whip hand, forced Daly to accept.

A third story, less widely circulated, is that the feud was not primarily a feud between Daly and Clark but between one of Daly's leading financial backers, James B. Haggin, and Clark. According to this account, Haggin, popularly believed to be of Turkish descent, had an engagement with Clark to which he was late. Clark, always punctual, waited irritably for a time and finally left, announcing that he would not wait for any "nigger." The remark was reported promptly to Haggin, and the feud was on, forced through Daly by his friend and backer.

In support of that theory, William Scallon, resident director and counsel for the Amalgamated in Butte in its first years of activity, and still practicing law in Helena, tells of an incident reported to him by Senator T. C. Powers. Powers, after Clark had once been denied a seat in the United States Senate on the grounds of bribery and had later won an election and been seated, commented facetiously to Haggin concerning the Clark success.

"He got there," said Powers.

"Not while I was interested in Montana," said Haggin.

Certainly the friendship between Daly and Haggin was close enough and enduring enough to indicate that Daly would have been willing and eager to fight Clark to the last ditch if Haggin desired it; whether or not he had any more personal motive for such a war. In proof of that friendship a story printed in the *Boulder Monitor* twenty-five years after Daly's death may be quoted:

"Harry Gallwey, who enjoyed the friendship and confidence of Marcus Daly from boyhood days in the Comstock region, once related a story as told him by Daly in connection with Haggin's loyal support.

"The Anaconda company had been wrestling with the problem of smelting copper. A plant was erected and found wanting. Copper smelting as far as refractory ores were concerned was a new industry. Experiments were expensive. At a meeting of the stockholders the matter of further expenditures was discussed. Tevis urged a stock assessment. Daly protested that his funds were exhausted, and he could not meet a levy. Tevis replied that in that case the only alternative would be to sell Daly out.

"Haggin, who had sat silent, interrupted. He drew a check book before him and began busily to sign checks in blank. When he had signed all of them he tossed the check book to Daly with the remark, 'Here, Mark, if you are broke, I'm not. You have always stood by me. I am going to stand by you. Take that check book and pay assessments as long as I have a dollar in the bank.'"

It is possible that all three of the stories cited as to the inception of that famous feud are true, one cause of friction leading to another. The battle reached epic proportions, with charges which could never be forgiven, in the contest over location of the state capital. There simply was not room in Montana for two such men as Daly and Clark without a battle. And such a battle!

Yet it was a fight in which Daly, at least, sometimes scorned to take advantage of an opportunity. John Gillie, veteran of more than half a century in the mines of Butte, for years chief engineer of the colossal Anaconda Mining Company, still their consulting engineer, is authority for one such incident.

The company owned a railroad, the Butte, Anaconda and Pacific, from its smelter at Anaconda to its mines on the hill. Clark owned the street railways in Butte, and a franchise for a line up the hill to Walkerville. When Clark started to build across the B. A. & P. tracks, Daly's manager of that line informed him that he would have to go into court to get permission. And in those days the courts were so crowded with mining suits that delays of months and years were common. Daly's manager reported to his principal with evident satisfaction that he had put a spoke in Clark's wheel. Thereupon Daly summoned Gillie, went to examine the crossing concerned, and promptly ordered Gillie to make arrangements for Clark to build his streetcar line crossing without the necessity of a suit, or the months or years of delay involved.

"That," says Mr. Gillie, who knew Marcus Daly intimately, socially and in business, throughout the last twenty-odd years of the copper magnate's life, "was typical of Mr. Daly."

In any event, the lines were drawn for the battle when Marcus Daly took over, for thirty thousand dollars, control of the Alice Mine at the top of Butte hill, and built a forty-stamp mill to work its ores. The deal was made for the Walker Brothers, with a small interest going to Daly personally. Butte's gold days were ended. Its silver days were come. Its copper ore, which eventually was to prove it "the richest hill on earth," was then hardly more than a nuisance in the extraction of its silver.

Shafts were sinking on a dozen claims. While most of the mines and mills were making but small profits because of the long and expensive haul of freight from the railroad at Corinne, Utah, or from Fort Benton on the Missouri River, they were providing a payroll mounting from the twenty men found by Clark in 1872 to many hundreds of men in the late 'seventies.

Marcus Daly found time not only to deepen and extend the workings of the Alice and to improve its milling facilities, but to examine neighboring shafts and outcroppings on the hill. Clark found time to bring the Original into profitable production, and to ship one hundred and fifty tons of ore to a newly constructed smelter near Central City, Colorado, in which he had a small interest. In the returns from that shipment was evidence that the copper which was making the silver ore difficult to work in the stamp mills might be of some value in itself if transportation costs could be eliminated by establishment of smelting facilities in Butte. Butte was a thriving town. It needed a bank. Clark had both the capital and the experience necessary. He opened a branch of his Deer Lodge bank under the firm name of Donnell, Clark & Larabie, later Clark & Larabie, still later W. A. Clark & Bro.

At the same time Marcus Daly's development of the Alice was proving the value of Clark's Moulton claim near by. The banker-miner could hold the undeveloped property with assurance that its value was increasing each day without the stroke of a pick or the expenditure of a dime. Everything was lovely. If and when he did decide to develop the Moulton its water would be likely to drain into the Alice workings below, and save him pumping costs.

But Marcus Daly had other ideas. He was not a man to work all his life for a salary any more than was Clark. Fifteen years in the mines and mills of the Comstock and in investigative work for Walker Brothers had made him a practical geologist of almost inspired discrimination. His observations in the first two hundred feet of depth attained in the Alice had inspired a vision—of a hill impregnated to a depth of thousands of feet with ore which in some veins would run almost pure copper.

Among the mines to which he had been granted admittance for inspection was one on the southeastern slope of the hill, owned by Michael Hickey, and named by Hickey the Anaconda. No story of Butte is complete without the explanation of that name. Hickey, according to this popular tale, had been a soldier in the Union Army. There he had read an editorial by Horace Greeley in which the famous editor declared that McClellan's Army would surround Lee's forces "like a giant anaconda." The name so impressed the Irishman that when he found his way to Butte and located a claim on the hillside he called it Anaconda.

Hickey admitted Daly to his shallow workings. Daly, already familiar with the ore and formation visible in the Alice and other shafts, here convinced himself that the big hill was a treasure house such as the world had never known, nor is likely to know again. Straightway he reported to the Walker brothers, urging purchase of the Anaconda. They sent another mining expert to make an examination. The second expert reported adversely. The Walkers declined to buy. Daly thereupon sold his fractional interest in the Alice for thirty thousand dollars and paid the money over to Hickey for control of the Anaconda. He was on his way to a fortune of as many millions.

But he lacked the money for development. He did not lack the confidence of mining men with ample funds. Greatest of these men in America at that time was George Hearst, with James B. Haggin and Lloyd Tevis as close seconds and associates. Daly had made the acquaintance of Hearst and Haggin in his Comstock days. In the summer of 1872 he had directed their attention to the Ontario Mine in Lake Flat, Utah. At the time it was hardly more than a prospect, available for purchase at thirty thousand dollars. And out of that prospect Hearst and Haggin were taking millions of dollars when Daly approached them with a report of another and greater prospect in the Anaconda at Butte. The Ontario produced seventeen million dollars in ten years, and paid dividends of six million two hundred and fifty thousand dollars.

Naturally Hearst, Haggin and Tevis were favorably disposed toward Marcus Daly. Hearst, himself a genius in geology, untrained as was Daly except in the practical schools of mine and mill, but with a similarly inspired nose for ore, had unlimited confidence in the Irishman.

It had been justified in the Ontario. Virtually ruined in the panic of '73, Hearst had sold everything he possessed to continue development of the Ontario, and had won his reward. By the time Daly's proposition concerning the Anaconda reached the Hearst-Haggin-Tevis group, Hearst had opened the Homestake Mine in Deadwood Gulch, South Dakota, the greatest gold mine in the history of the world, and was paying himself and his partners dividends. The Homestake was to produce eighty million dollars in twenty years, and to continue indefinitely pouring its vast wealth into the Hearst family coffers.

That apparently inexhaustible stream of gold, together with the vast profits from the Ontario, made it an easy matter for

Hearst, Haggin and Tevis to finance Marcus Daly's dream. It is even reported that Hearst made his investment without looking over the property. Inasmuch as the Hearst-Haggin partnership operated on two fundamental rules, (1) never to buy an interest in property controlled by others, (2) never to pay more than the value of the ore in sight, the truth of that report seems doubtful. There was little ore in sight in the Anaconda. But there was Marcus Daly's word for it that a hundred million dollars' worth of ore lay below. Probably Hearst came, examined the property and the hill, and approved.

In any event the Anaconda was soon on its way, with Daly in charge, backed by the millions of Hearst, Haggin and Tevis, and the vast resources of the great Homestake and Ontario Mines.

In the meantime, Clark had aroused the interest of the Black Hawk smelter management near Central City, Colorado, to which he had shipped sample ores from the Original. They organized the Colorado & Montana Smelting Company and built a smelter, partly with Clark money, on the south side of Silver Bow Creek, below the town of Butte. The smelter operated successfully, returning a profit to Clark on his investment. Another profit came through the more economical treatment of his ores, and a third profit to Butte in general through the purchase of ores of various mines necessary to obtain the proper fluxing mixture for efficient operation of the smelter.

The town was busy, prosperous, growing and happy when its attention was drawn abruptly from its business affairs to sudden realization that it was still a frontier town, far from railroads, far from the safety and refinements of civilization. News that the Nez Percé Indians under their famous Chief Joseph had gone on the war-path was brought into Butte and the smaller towns of western Montana by excited riders on galloping lathered horses.

The fact that the reports of danger, the accompanying tales of atrocities by the Indians, were exaggerated was not even suspected in Butte or Missoula or Deer Lodge. The famous Custer massacre had occurred only a year earlier. There were scores of men in the region who had actually seen the horribly mutilated bodies of Custer's command lying upon the battlefield.

When William A. Clark galloped down from Deer Lodge to Butte, thirty-odd miles by the shortest trails, in three hours, bearing a telegraphic commission from Governor Potts to raise a battalion against the Indians, Butte sprang to arms. Three companies of volunteers were formed before nightfall, and, under Clark as major, set out to intercept the Indians, who were reported to have crossed the Big Hole River near its source. The volunteers traveled all night and most of the next day, but before they reached the designated point the Indians had clashed with a small command of federal troops under General Gibbon. Seventy-five white troopers were killed or wounded in that battle.

But Chief Joseph was a strategist of high ability. He was not primarily upon the war-path at all, but was trying to lead his people, including squaws and children from the Idaho reservation upon which they had been virtually interned by the encroaching whites, to a new home in Canada. It was not his purpose to kill, but to escape. The story of the so-called Nez Percé war is one of the blackest marks in the whole black record of American "civilization's" treatment of the savages. But Butte was not analytical. It sallied forth to battle, and only the fact that Joseph had fled after his defeat of General Gibbon prevented the motley assortment of volunteers, miners, gamblers and merchants, from contributing their blood to the cause. The Indians fled to the Snake River country and were eventually rounded up by General Miles.

Butte congratulated itself upon its bravery, and returned to its labors. Its future was assured. The population doubled in the five years between Marcus Daly's arrival in 1876, and the advent of its first railroad, the Utah Northern, which connected the town with Salt Lake City in December, 1881. It was to double again in the next five years with the building of additional mills, smelters and concentrators, the opening of scores of mines, and the completion of its second railroad, the Montana Union, a spur of the Northern Pacific. It was to double again by the time its third railroad, the Montana Central, reached it in the summer of 1888.

Those were the days of which the remaining pioneers of the town, some still in business or professions there, others scattered from New York to Los Angeles, delight to tell. Talking with some of them, seventy, eighty years old, is a revelation. The deadly air of Butte, thick with the fumes of sulphur, arsenic and smoke from the open roasting of its ores and from the stacks of its smelters, killed every blade of grass, every flower and every tree within a radius of miles. It could not kill such men as Lee Mantle, John Gillie, Jerry Clifford, John J. McHatton, W. I. Lippincott, Charles Leonard, Alex Johnston, John Hauswirth, Ben Stack, William Scallon and numerous others who have searched their memories to contribute to this record. Those were days of sweat and struggle, low living, high living, great accomplishment, a devilish sense of humor. Anecdotes never adequately recorded except in the memories of the pioneers are revealing.

CHAPTER V

THE SPIRIT AND THE FLESH

PROBABLY there were more kinds of people, more character types, more divergent interests in Butte in the early 'eighties than in any city of its size in the world. Marcus Daly brought in shiploads of Irishmen to work his properties. Clark preferred the Cornishmen, popularly known as Cousin Jacks, efficient with the inheritance of generations of miners. Veterans of both North and South in the Civil War were there in numbers. There is still a legend current in Butte that Price's famous lost regiment, never mustered out of service, marched into Montana when the Confederate cause was lost, and started life anew in Butte. Unquestionably, as revealed in the instance of Michael Hickey, of the Anaconda, discharged Union soldiers were early settlers there.

The big bonanza of the Comstock Lode had just paid its final dividends of one hundred and five million dollars to Mackay, Fair, Flood and O'Brien, and Virginia City, Nevada, had collapsed when Butte discovered that it was rich in silver ore. Comstock miners flocked to Butte. Business men, gamblers, dance-hall girls, and all the motley rabble of mining-camp followers crowded in. Despite that conglomeration, the city, being busy and prosperous, was reasonably honest and law-abiding with reference to property, though its morals and its sense of humor were unrestrained.

John Gillie, who arrived in 1880, and is still one of the most respected residents of the state, is authority for the statement that in those exuberant days petty crime was almost unknown. For two years he lived in a tent, fastened only by a loop of tape, and though frequently absent for days at a time on various mining missions, never lost one item of property.

Those Utopian days were brief. Within a month after the arrival of the first railroad, in December, 1881, the town was crowded with tramps, thieves, burglars and the riffraff of humanity. Burglaries and stickups on the poorly lighted streets occurred nightly. It was the period in which the depredations of the Jesse James gang had brought that outlaw into national notoriety. Whether he actually came to Butte, or his name and fame were used by another desperado to promote his own crimes is not clear from the casual records, although there are still men in Butte who will name the hotel where he stopped, and insist that he was constructing a tunnel to rob one of the banks when driven out by a shooting scrape.

It is certain, however, that one "Tex" Edwards, a widely known desperado, approached J. A. Kennicott, clerk in the bank of Donnell, Clark and Larabie, and offered him inducements to join the Jesse James gang in a plot to rob that institution. Kennicott was believed to know the combination of the vault. He, knowing Edwards' reputation, asked time to think the proposition over, and went straightway to J. Ross Clark, brother of the bank's founder, and suggested plans to thwart the robbery.

One of Edwards' associates, possibly a member of the Jesse James gang, trailed Kennicott and reported to Edwards. The bad-man, meeting Kennicott next day in King & Lowry's saloon, accused him of giving away the scheme and warning the bank. To back his abuse, "Tex" reached for his gun. But Kennicott was more than a bank clerk. He was handy with a gun himself, and before "Tex" could complete his draw, shot him through the throat, killing him almost instantly.

Threats by the criminal element, directed alike against local authorities and private citizens soon grew alarming. But the honest residents of Montana had not forgotten another January, eighteen years earlier, in which similar threats and crimes had been answered by the peremptory hanging of twenty-one desperadoes. Neither had they forgotten the symbol used by that early vigilance committee on all communications directed against the organized criminals—the figures "3-7-77."

A group of citizens met in Henry Jacob's store on Main Street, and before the night was over notices were posted all about the town, warning all thieves, footpads, burglars and other criminals to leave town within twelve hours or face the penalty of summary execution. The notices were signed "3-7-77."

The bad-men knew what that meant. They promptly took to their heels. So many of them arrived in a group at Deer Lodge that the citizens of that peaceful community, forewarned by telegraph, met them with rifles and shotguns and routed them around the town. Butte was made at least temporarily safe for life and property. Even the local courts were impressed.

When Judge DeWitt dropped in one day to inspect the home of his brother-in-law, Harry D'Auschel, upon which he had been asked to keep an eye during the temporary absence of the family, he found all the furnishings except the piano missing. In the basement, rolled in rugs taken from upper floors, were a few articles, evidently left for later removal. DeWitt engaged Sam Trice, formerly a police officer, to watch for the return of the burglars that night.

Some time after midnight two skulking figures appeared at the back of the house, raised the old-fashioned doors which closed down upon the outside stairs of the cellar, and descended. When the first man reappeared with a bundle on his head, Trice fired a double-barrel charge of buckshot, literally blowing the man's head out from under the bundle, which dropped upon the burglar's shoulders before the body collapsed.

The second burglar promptly surrendered. Judge DeWitt was county attorney at the time and presented the evidence to the grand jury that very morning. Within an hour a true bill was returned, the remaining burglar haled into court, and after a plea of guilty was sentenced to the territorial prison at Deer Lodge, and started to that destination.

That was the manner in which courts and citizens operated when necessity arose. They were not so particular with reference to personal morals. John Gillie, with a twinkle in eyes which show evidence of seventy-odd years of laughter, throws light upon another phase of Butte life as revealed after the completion of its second railroad.

When the Montana Union connected Butte with the Northern Pacific at Garrison in 1883, the junction platform quickly became a rendezvous for neighboring Indians, begging or selling baskets and bead-work to travelers changing trains. On one occasion a squaw, with a solemn-faced papoose strapped to her back, attracted the attention of a woman traveling to Butte for the first time. The woman was interested in the baby, which appeared to be of lighter complexion than most papooses. She addressed the mother on the subject, speaking patronizingly in what she supposed to be the vernacular of the Indians. "Him full-blooded Injun?" she asked, pointing to the child. "No," the squaw answered after a moment, her eyes expressionless, "him part Injun, part Injuneer."

Butte was not what one might call discriminating, except in the matter of mines. At one time, before the corrosive effects of sulphur and arsenic destroyed it, according to one of the elderly wags still living, there stood in the cemetery a wooden headboard bearing this epitaph: "Here lies the body of Martha Goff. She had two sore legs and a badish cough. It wasn't her cough as carried her off, but it was a coffin they carried her off in." Better substantiated is an obituary notice printed in one of the town's first newspapers, which ended with this sentence: "He frequented the Silver Bow Brewery with faithful regularity for the last two years of his life."

So much for Butte's sense of humor. It also had its sense of dignity, occasionally revealed in most illuminating language in police court reports. For instance, this unique judicial utterance under date of June 10, 1880, evidently by a police court justice who had been summarily thrown out of office:

"To Whom It May Concern: Know ye, that the undersigned this day casts off the judicial ermine and retires to the walks of private life, with a reputation that does not meet with the approval of the incoming administration. It has been our aim during our brief career as the chief justice of this city to perform our duties in an impartial, upright and fearless manner, and had flattered ourself that we had accomplished our aim. But evidently we do not see ourself as others see us. The inaugural address of His Royal Highness, Mayor Valiton, bids us leave all hope behind of further serving the city in the capacity of an official. The Jeffersonian test for office, in the language of the keeper of the coal yard, 'I hope you will confirm nominations for the several offices. I will appoint men well known for their honesty and integrity.'

"Those who have grazed in the pastures beside me for the past few months, being well known for their 'honesty and integrity,' will for a few brief moons continue installed at the crib, when

"'They, like the flower and the weed, will wither away to let others succeed.'"

It seems clear that whatever the justice's judicial and political qualifications might have been, he was a man of sonorous language. In contrast we have a police reporter's account of an incident in Judge McHatton's court: "A ranch employe[e] named Regan came to town yesterday and got full, resulting in him being run into the cooler. This morning he was arraigned before Judge McHatton, who fined him $5 and costs. Thereupon the bucolic bum got saucy, told the court that he'd give $5 and not a——cent more, and wound up with an indecent remark, which is not fit for repetition. The words were hardly out of his mouth before the court fined him $100 for contempt and sent him to the dungeon cell for 24 hours. It won't do to monkey with Judge McHatton's court."

Neither did it do to monkey with some of the men in private life in those days. Gunplay was not frequent, as gunplay is expected to go in mining camps, but occasionally it was furious. Outstanding in the records is the historic battle between "Bodie Joe" Rowells and "Lean Joe" Campbell, which gave the residents of the town something to discuss over their Christmas dinners in 1881.

Ben Stack says the crowd in King and Lowry's saloon and gambling house, where the battle occurred on Christmas Eve, tore out the whole front of the building in their haste to escape from the flying bullets. Judge W. I. Lippincott, who also was there and who is not given to understatement in reference to dramatic incidents, says they tore out the double front doors. The judge, then a youth with aspirations toward the practice of law, watched the duel from a comparatively safe vantage point behind the great iron stove. Here is his story:

"I walked into King and Lowry's gambling house to see what was doing. At the first faro table sat Davey Allen, dealing, Jakey Thompson in the lookout chair, and old Dick keeping cases.

Saul Millen and a lot of the old ones were playing bank. Behind the bar was George Odell, waiting on customers, and Jim Murray near the safe. Mel Lowry and Silas King were in and out and there was quite a crowd in the saloon. 'Pegleg' was near the restaurant in the rear of the long room, dealing stud poker, and everybody was busy.

"Long 'Lean Joe' Campbell had been having some trouble with 'Bodie Joe.' Campbell said: 'I just bet I'll have trouble with "Bodie Joe," and I just loaned my good gun. I'll be in a fix if he comes as he said he would.'

"There were half a dozen men waiting around and listening to his talk. He had an old short-barreled white-handled gun of large bore. I stood close to Campbell, and he was leaning on the bar. In a moment in stepped 'Bodie Joe' with a man, and said, 'There is the son-of-a-bitch,' and stepped back to the restaurant, and came back at once with a gun in his hand. The bar was crowded and the room was crowded. 'Bodie Joe' shoved himself between a couple of men, and Joe Campbell said, 'Don't shoot here. Come outside and I'll give you what you want.'

"'Bodie Joe' says, 'I'll give it to you right now,' and fired. The first shot hit Campbell between the thumb and index finger, and all the crowd that could get past ran for the big double doors in front, broke them down and carried them into the street. I jumped behind the big stove and watched the duel. The next shot from 'Bodie Joe' went wild and went across the street, hitting a man named Martin, just above the hip, the bullet running around just under the skin and stopping just above the other hip. Doctor Thompson cut it out and pronounced the wound nothing serious. Martin was a teamster hauling ore for the Bell Mine, then being operated by Charles T. Meader, with Alex Talbott for superintendent. Incidentally, Alex Talbott at that time was the youngest mining superintendent we ever heard of, and probably the youngest in the world, but he was a good one and made good a-plenty.

"Well, the next shot hit Campbell on top of the head and glanced and struck a large mirror over the back-bar, and shattered that. The next shot struck Campbell on the lower jaw; the next on the right arm, above the elbow, shattering the bone.

"Campbell had all this time been trying to get his gun out of his right hip pocket with his left hand, his right having been disabled by the first shot. Also it was an old gun, and strange to him. But he finally got it out, stepped forward, shoved his gun against 'Bodie Joe's' stomach and fired, using his left hand. "'Bodie Joe' turned and ran for the restaurant, but Campbell had got him. Campbell didn't know that, and ran after him, trying at the same time to get a knife open with his teeth. But 'Bodie Joe' fell, and we picked him up and laid him on a faro table. Bodie says, 'He's got me, boys, but someone will get him.' He was taken to St. James Hospital and died in a few minutes, with only one bullet in him, and after emptying his gun into Joe Campbell. The last moments of 'Bodie Joe' were terrible to hear, for he swore at his luck and everything in sight until death cut off his abuse and profanity."

A coroner's jury promptly exonerated Campbell, but he demanded an indictment and trial as soon as he was able to leave the hospital where his wounds were treated. Although he had been wounded twice in the head, once in the arm and once in the hand before he had fired a shot, he was around his old haunts in two months. When he was brought to trial the jury acquitted him without leaving their seats. Later he became warden of the Idaho state penitentiary, and was long a respected citizen of that state. There were hardy men in Butte, and resourceful. Witness the report of a surgical operation as described by Mrs. Fogarty, formerly of the Butte

Public Library, in notes now preserved in the State Historical Library:

"Mr. John Gill is best remembered by the manner of his death. He was found one day in an unconscious condition with his skull fractured, and a piece of the bone pressing against the brain. It looked very much as though he had been hit with the iron end of a shovel. E. H. Butler testified that he had fallen over an embankment and hit his head against a rock. His friends thought that if they could bring him to consciousness he would testify that it was done with a shovel, for they suspected Butler of the murder. Doctor Beal did what he could for the man but there was not a surgical instrument in the town, and he was forced to use what came to his hand.

"The man being unconscious they were able to perform an operation without an anesthetic. A bit and brace were used to gain a hold on the piece of skull and bring it into place. Mr. A. W. Barnard held the man's head while the operation was performed, and he remembers to this day the feeling of horror and apprehension lest the instrument go too far.

"But the operation was entirely successful, the bit of bone was raised, and the patient opened his eyes and looked around, but volunteered no information. In order that the bone could be held in place for a time, Mrs. Perks gave a silver ring which was cut open and flattened out, and one end inserted under the bone to pry it up. But even then it was necessary for someone to hold it all the time. After a number of hours, the man never being conscious enough to enlighten them, they gave it up and he died."

Such was life and death in Butte while its silver production, its mills, its smelter facilities, its transportation facilities, its population and its general prosperity mounted steadily.

Marcus Daly had opened the Anaconda and mined and milled far more than enough silver to pay himself and his associates,

PHOTOS COURTESY OF MONTANA HISTORICAL SOCIETY, HELENA

Right: *Marcus Daly, called by a Butte newspaper at the time of his death "Greater than Napoleon."*

Above: *The 1882 discovery of a huge copper body in Marcus Daly's Anaconda Mine set the scene for the greatest mining camp on earth.*

Below: *One of the stories about the beginning of the Clark-Daly feud centers around Daly's interest in the Alice Mine in Walkerville in 1876.*

Left: *W. A. Clark celebrates his 1894 victory over Marcus Daly, a battle which made Helena the state capital and cost each side an estimated $1 million.*

Right: *William Andrews Clark worked sixteen- and eighteen-hour days to keep every detail of his widespread business in his own control.*

Below: *W. A. Clark's Colusa Smelter in Meaderville, 1889. Clark started working the copper deposits of the Colusa in 1873–1874. He was the first to ship commercial quantities, which had to be hauled four hundred miles to Corinne, Utah.*

PHOTOS COURTESY OF MONTANA HISTORICAL SOCIETY, HELENA

Haggin, Hearst and Tevis, a profit of several hundred percent on their investment. But the silver ore was almost exhausted at a depth of three hundred feet. That did not alarm the Irish immigrant who had been inspired with a vision of empire-building proportions. He had seen the stains of copper within the hill. When men of lesser vision saw the increasing copper in the diminishing silver ore, they were panic-stricken at the thought that their rich silver properties would soon be exhausted. Not Marcus Daly. He realized that here was greater wealth than anything conceived by the grubbing independent silver miners of Butte, or even by the shrewd and calculating William Andrews Clark.

When the independents had, for the most part, reached the sad conclusion that Butte's great days were over with the decrease in its silver production, Daly closed down the Anaconda. The valuation of neighboring properties dropped swiftly. Daly carried his information and his vision to his associates. They, with millions already gained through Daly's tip on the Ontario, and considerable profit already taken from the shallow shafts of the Anaconda, the St. Lawrence and the Neversweat, were easy to convince. They gave Daly full power and sufficient capital to go ahead. He bought surrounding claims at bargain prices. Then he reopened the Anaconda and announced to the world that Butte was "the richest hill on earth," the center of the world's future supply of copper. And the world, with the development of electric power and light, was beginning to realize that copper was one of its most important requirements.

Daly had proved to himself and his partners both the extraordinary richness of the copper ores and the fact that they could be economically treated. This he had accomplished by shipping tons of the rock for smelting and analysis to almost every smelter then in successful operation. Some ore had been shipped as far as Swansea, Wales. It remained now to put the technical information so gained into effect in a smelter close to the mines.

William A. Clark, with the added advantage of his scientific knowledge of metallurgy gained in the Columbia School of Mines, and improved in the practical workings of the Colorado and Montana Smelting Company, now reorganized that concern with himself in control. Thereupon he installed new equipment and a concentrating plant across Silver Bow Creek from the town and mines.

Daly's vision was broader, grander. While Clark planned to gather a vast personal fortune, Daly planned to build the western territory of Montana into an empire of untold riches coming from the Butte hill.

The lines were drawn for battle.

CHAPTER VI

THE DEVIL'S THRONE

BUTTE meanwhile had grown far beyond the few scattered huts, log houses and more or less imposing Hotel de Mineral—of two stories in one section—to an almost solidly built area four or five blocks square, with some scattered miners' houses. Within a couple of miles were the politically independent settlements of Meaderville, Centerville and Walkerville. This had been accomplished with the operation of its first successful silver mills, from 1878 through the next few years. From the business center on the lower southeastern slope it was extending up the hill, and down toward the flat. Town Gulch, where the first houses had been built among stunted pines, had become Dublin Gulch, because of its predominating population of Irish.

New processes of treating the refractory ores from its deepening mines included the open roasting of these ores with a mixture of salt to eliminate their sulphur and arsenic content, convert the silver into chloride and render the base metals recoverable at a profit. The sulphur and arsenic fumes with the other smoke of the burning ore heaps swept the town, the hillside and the surrounding country.

Grass and flowers died at their first touch. The trees which had not already been cut down for fuel or building purposes dropped their needles and leaves, and died. Cats licked the all-pervading grime from their whiskers and were poisoned by the arsenic it contained. Cattle, grazing on the withering grass at a distance of miles, seemed literally to plate their teeth with copper. That is vouched for by Isaac Edinger, still a rancher thirty miles south of Butte, who purchased thousands of the cattle for slaughter.

"I used to carry a few of those gold-colored teeth in my pocket all the time," he says, "because no one would believe me, and I'd have to show 'em. When they were shown they always wanted to keep the evidence, and I'd have to get a new supply every time I went back to the slaughter house."

On a windless day when a low ceiling of clouds hung over hill and city, holding down the smoke, lamps burned in the houses and upon the streets at midday. Even then there were times when a man could hardly see his own hand through the brown and stinking murk. Burglars, stickup men and sneak thieves did not have to wait for night to pursue their calling.

In illustration is a story about Judge John McHatton, standing upon the steps of the court-house at ten o'clock in the morning chatting with friends. It is said the judge was robbed of a diamond pin and watch, in what should have been broad daylight, with Sheriff Lloyd standing ten feet away, and the pickpocket escaped in the darkness.

"It must have been some other judge," says McHatton. "It wasn't I. But the interesting thing is that such an incident could have been true. It was that dark in Butte at times. I always carried a gun."

That was Butte by the time it reached a population of fifteen thousand—barren, sulphurous, smoke-ridden. Only man could survive such an atmosphere. Only man, held by Marcus Daly's assertion that this was "the richest hill on earth" would attempt it. Even some of the men, and more of the women, worn by their never-ending battle against the soot which ruined their clothing, begrimed their floors and festooned their curtains, scoffed at the phrase. "The richest hill on earth. Huh! It's the perch of the devil." So a second subtitle was given to Butte.

But Butte could survive the second because the first was proving true. Men and women will defy or connive with the devil for riches, and at times even for a livelihood. For profit they will live and labor literally in an atmosphere of hell, breathing brimstone, festooned with soot, surrounded by the imps of Satan, and, such is their heritage, find time and courage to laugh and to play. Easy money makes a wild town. Butte had it, and was it. Few of the prominent early residents took life as seriously as William A. Clark. Even such important men as Marcus Daly, Lee Mantle, Pat Largey, John Noyes and A. J. Davis, with a keen eye to business, had another eye for something else. Life in Butte might be sulphurous, but it was hilarious.

When W. I. Lippincott, J. H. Montieth and Richard A. Butler, popularly known as "Comical Dick," started the *Mining Journal* in the early 'eighties, Henry T. Meader was just opening the Modoc Mine. Meader, whose name is still preserved in Meaderville, center of Butte's night life in 1934, had money, but needed more. Publicity is the life blood of promotions. Meader came to the *Journal* and offered one thousand dollars a column for a write-up of the riches revealed in the Modoc workings. The young editors were happy to oblige.

Six columns of print were wrangled out of the shallow shafts and promises of the Modoc. Meader paid the six thousand dollars. Immediately the *Journal* staff went out to celebrate. By the time the celebration had burned itself and the six thousand dollars out, several issues of the paper had failed to appear. Its editors were too busy. Incidentally their interests had been turned into other fields. Opportunities of all kinds abounded in Butte. E. W. Penrose took over the paper and ran it until he was killed by three men who took exception to some of his comments on life and character in Butte. The killers were never tried, according to Judge Lippincott, who is likewise authority for the earlier incidents of the story.

The devil was enthroned upon the richest hill on earth. Action was inevitable.

With the completion of its first railroad, Butte was a city. It was no longer necessary to haul its refractory ores over hundreds of miles of wagon roads, and ship them to smelters. It was no longer necessary to bring its supplies and luxuries in over the same difficult and costly route. Its streets had been improved with its houses. It was no longer necessary for teams passing along its Broadway to detour almost down to the flat to cross the gulch which gashed the hillside back of the Hotel de Mineral. The gulch had been bridged. Soon it was to be filled and called Hamilton Street. It was no longer necessary for residents of the community to go to one of the three hydrants which provided the town's water supply in slightly earlier days, and carry home a bucket.

The Silver Bow Water Company, promoted and owned largely by John Noyes, O. P. Blaine and William A. Clark, was piping water from Bull Run Springs, five miles away, into half the homes and business houses. There was material evidence of the advance of civilization. Incidentally, with Clark's money in the enterprise, there was evidence of the substantial promise of the region. Clark made few economic errors. Homes and business houses were improving in architectural quality as well as in numbers while the gardens and trees withered under the blighting smoke which helped to bring about that improvement.

Butte was the richest and most populous town north of Salt Lake City and between St. Paul and Portland, an area nearly fifteen hundred miles long by six hundred miles wide. It could no longer submit to dictation from Deer Lodge, the distant seat of its county government. It proclaimed its independence, making itself the seat of a new county, Silver Bow, cut off from the southern end of Deer Lodge County.

In the next year Butte netted $6,300,000 from its ores, mostly in silver. In the following year production was boosted to $8,928,000. Attracted by such wealth and promise of growing freight and passenger traffic, a branch of the Northern Pacific built into Butte and a new wave of prosperity flowed over the territory. The next year the Butte mines produced $16,000,000, still mostly silver, and new stamp mills were added to treat the ores.

In the meantime Marcus Daly had sunk his shaft on the Anaconda through one hundred feet of low-grade silver ore into a seam of "copper glance" that carried thirty percent of the red metal. He followed down the seam to a depth of three hundred feet. And there the ore body widened to five feet. Daly had proved the accuracy of his vision. He had stood upon an upthrust ledge of rock, looked three hundred feet into the hill, and told his friends that here would be the greatest copper mine the world had ever known. He had opened the hill so that they also might see. They saw, and were convinced.

It remained to build a smelter large enough to treat the vast tonnage of ore which had been proved. The Parrott smelting works were already treating copper ores taken from the Parrott vein. The Colorado and Montana Smelting Company, reorganized under Clark's control, was prepared to make certain that Clark lost none of the wealth of his Travona, Moulton, Original, Colusa and other properties. But those smelters were small compared with the vision of Marcus Daly.

He sought a site adequate for the greatest copper smelting plant in the world, and found it near Warm Springs Creek, twenty-six miles from Butte. There he laid out the town of Anaconda, and prepared to build his smelter. Money was easy to get. The plans were in keeping with the vision.

Copper mining, since the ancient days when the Carthaginians whipped their ore-carrying slaves up the long ladders in the mines of the Rio Tinto, has been a business requiring large capital and extensive organization. A copper mine, to be anything, must be vast. Millions must be invested in both mining and refining equipment. To attract such capital, the evidence must be clear of an ore body sufficient to justify great expenditure through many years. Marcus Daly had made it clear in the Anaconda. Hearst, Haggin and Tevis had the money. The first Anaconda smelter was built, with a city rising simultaneously below it.

The attention of wealthy men throughout the country, already attracted to Butte, was concentrated. The city of Boston then, by various turns of events, was the center of the copper trade and financing of America, if not of the world. Boston and Pittsburgh had financed the Calumet and Hecla properties on the south shore of Lake Superior, and had taken millions of dollars in dividends from them. Copper dividends to the amount of millions annually were flowing from Calumet and Hecla into Boston when Butte began to promise competition. Boston bankers, brokers and business men were copper-conscious. They could afford to be.

A. S. Bigelow, J. W. Clark, who was no kin of William A. Clark, and other Bostonians who had taken fortunes from the Lake Superior mines, formed the Boston and Montana Company, and later the Butte and Boston Company to take more millions from the new copper district. Plenty of money was available to develop the mines of Butte. At first the capital came out of the hill itself, but soon it was coming from some of the wealthiest men in the United States. They were eager to invest in a business which promised, and, in the course of years, paid scores of millions in legitimate dividends, and even more millions in stock-market operations far less honest.

Daly started the great Anaconda smelter in 1883, while the city of Butte was still financing local business largely with the silver taken from its mines, though building high hopes upon the copper which lay below. Daly was rough, genial, unassuming, generous, popular. He still maintained an unpretentious home upon the hillside not far from the Anaconda. William A. Clark, in contrast, was dapper, immaculate, polished, polite, inscrutable, a character almost unique among the rough diamonds of the American frontier which had bred and trained him. He kept his home in the more cleanly, tree-shaded and garden-decked community of Deer Lodge, although he was building a pretentious house in Butte.

Deer Lodge was still headquarters of Clark's banking and mercantile business, but he was now spending most of his time and nearly all his tremendous energy in his branch bank in Butte, in his mines and reduction plants, in the development of his street railway, and in the direction of his newspaper, the *Butte Miner.*

The mills designed primarily for treating the silver ores had grown in number until two hundred and ninety stamps were dropping, shaking the smoke-laden air of the hillside with a never-ceasing rumble. Business, since two railroads had reached the town and money for investment was pouring in from Boston and other sources, was booming.

More than three thousand miners were at work within the hill. Another thousand men were employed upon the surface, in mills, offices, stores and elsewhere. Common labor earned three dollars per day; miners, three dollars and fifty cents; timbermen and others more skilled, four to six dollars. The total payroll in Butte was nearing half a million dollars per month, and almost every cent of it was being spent within the city limits.

Efficient, energetic and discriminating men besides William A. Clark were already accumulating fortunes. The annual edition of the semi-weekly *Inter-Mountain* rated A. J. Davis, banker, real estate and mining operator, as the wealthiest man in Butte in 1884, with a fortune estimated at five million dollars. And Davis had been broke in 1863 when he arrived in Montana from Iowa. Such were the opportunities in Butte. Even William A. Clark was not rated so high in 1884. The same edition of the *Inter-Mountain* asserted that Clark paid on an assessment of three hundred thousand dollars in Silver Bow County in that year. His mercantile firm of Donnell, Clark and Larabie paid only two hundred and forty-three dollars and eighty cents to the Silver Bow tax collector, and his branch bank paid twenty-five hundred dollars.

Other men who took a leading part in the building of Butte were listed under a heading of "Our Heavy Taxpayers." Patrick A. Largey, banker and merchant, had to dig up $539. Marcus Daly was well down the list, paying $188. Lee Mantle trailed with $127. The Alice Mine was the richest producer on the hill. Its output totaled $1,300,000, and it paid a county tax of $5,324. The Anaconda, destined to become the greatest copper mine in the world, paid less than half that sum.

Still Butte was becoming interested in numerous things besides money. An opera house was promoted, and the city began to gain fame among troupers as one on the long journey from coast to coast where they were always certain of enthusiastic welcome, a full house and a cordial personal hospitality. Emma Abbott, Maud Granger, Adelaide Randall, Frederick Warde, George Milne, Dan Sully, Mademoiselle Rhea and other stars introduced a new and thrilling form of entertainment to frontiersmen who had never seen an artistic theatrical performance. Residents with a more cultivated background who had been starved for music and the theater were even more eager to escape the blighting smoke of their workaday world and refresh themselves in the romance of the stage.

When the first theater built by John Maguire was destroyed by fire, all Butte mourned. But it was not content to mourn. Not Butte. Within a day plans were under way for a program which would finance a new and finer theater. There was no place within the city large enough to hold the crowds which were eager to contribute. The race track was selected for the party. Half the women in the city baked cakes and made sandwiches. Every professional and amateur entertainer of note within a day's ride volunteered to perform. Miners came from the shafts, clerks from the stores, women from the tenderloin to pay their fees for the entertainment and buy the goods provided.

The theater was rebuilt. Butte lived again. Other theaters followed. There appeared, season after season, with profit to themselves and delight to their audiences, such artists as Modjeska, Melbourne, MacDowell, the Bostonians, the Carleton Opera Company, Ward and James, Minnie Maddern Fiske, John Drew, Bernhardt, Sothern, DeWolf Hopper, not to mention scores of others of a day when the stars of Broadway were willing troupers, as happy in the unstinted applause of Butte as of Manhattan.

For persons who preferred a theatrical diet less delicate, such houses as the Casino and the Comique were opened. They were typical of frontier towns and mining camps from the first rude entertainment of San Francisco's Barbary Coast. They catered to the lowest tastes of the town, and Butte's tastes were as depraved as they were cultivated. Boast as it might, and did, of its love of good music and dramatic artistry, Butte was supporting thousands of men and women whose idea of entertainment was crude, to say the least.

The Comique was designed for their entertainment, with admirable arrangements for keeping secret the identity of its theoretically respectable patrons. Tinhorn gamblers, pimps and others who had no reputations to lose, walked boldly in to the main floor of the "theater," seated themselves at tables and ordered drinks from the circulating girls in gaudy low-cut evening dresses. Then they helped the girls deposit their percentage tickets in their stockings, and applauded the vaudeville and burlesque which was presented on the stage.

Occasionally this class of patronage was supplemented by tourists or slumming parties of the more superficially respectable residents of Butte who felt that by entering boldly and merrily they could laugh off any implication of depraved tastes. For the vaudeville, frequently supplied by the leading booking offices of New York, was often excellent. Eddie Foy was one of the famous troupers who played there. When hilarious patrons lifted one of the percentage girls to a table top or to the stage and called for song and dance it was not so good, but its informality was accepted in compensation. That was the spirit and the letter of the main floor of the Comique.

For those who wished no spectators and no interference in their acts of helping the painted girls deposit their percentage tickets or a gold piece in a safe place next the skin, there was a row of screened boxes above the floor, to which entry could be gained by a stairway leading from the alley. Some of the city's leading citizens could, and did, come directly from a performance of the Bostonians, Mrs. Fiske or John Drew to these screened boxes.

Each box, entered from a narrow corridor invisible from the house, was provided with a bolt on the inside of its door and a slide through which trays of drinks or iced buckets of champagne could be passed. The occupants of the boxes could watch the stage show through a painted screen which guarded their own identity, or could entertain themselves more personally with so-called actresses summoned from stage or floor. The girls were always willing to oblige. If they got nothing more they were always rewarded with a commission on every drink they induced a patron to buy, a red percentage ticket being passed in through the slide with each order.

For still lower tastes there was a segregated district with houses ranging from some with palatial appointments to the most miserable cribs. As time went on and wealth increased the fame of some of the "madams" of Butte spread from coast to coast, rivaling that of the Everleighs of Chicago and Dolly Ogden of San Francisco.

There are few oldsters of Butte, whether living there today or in Boston, New York or Los Angeles, who cannot and will not, on occasion, tell you of the beauty, the hospitality, the charity, the high heart and low morals of Lou Harpell, Mabel Loy, Belle Rhodes and Molly Demurska. A favorite tale of the earlier days, related to the writer by the judge who officiated, concerns the wedding of Molly Demurska and Jack Jolly, a town marshal of Butte.

It was a wedding attended by a score of the leading men of the city. For the sake of their present conventional respectability it might be added that most of them were bachelors at that time, too busy with the engineering of their own fortunes to give much thought to domestic felicity. At least one of the guests was to become a United States senator. Others were to rise high with the building of Butte. For the moment they were at play.

Jack Jolly was a popular man who had suffered a tragedy which stirred the sympathy of the entire town. His wife and seven children, according to the judge's tale, had died within a few days in an epidemic of influenza, or as it was then called, la grippe. Jack had broken under the tragedy, sold off his property and run amuck with liquor and gambling. Friends obtained for him the post of marshal, and he straightened up enough to serve with fair efficiency.

When he announced to these friends that he was going to marry Molly Demurska, they planned a ceremony suitable to his regeneration. Molly was popular. She ran an honest house, beautifully appointed, with beautiful girls as her boarders. A select list of guests was arranged. The grand staircase of the house was decorated with smilax and flowers. A wedding bell hung at its foot for the ceremony. Hampers of champagne and the finest delicacies obtainable in Chicago and San Francisco were provided for the wedding supper.

The officiating officer himself prepared the bride for the ceremony. The girls of the house, in decollete which left nothing to the imagination, lined the stairway. The guests were assembled below. Jolly and Molly descended the staircase to the strains of Mendelsohn's *Wedding March,* played by the "professor." The judge pronounced them man and wife. The bridesmaids stood on their heads with delight. Molly Jolly was an honest woman. Her last indiscretion had been with the judge.

But Fate was unkind. The bride and bridegroom set out for Alaska and Jack Jolly killed a man on the train in a gambling row. Shortly thereafter he himself was killed by Soapy Smith, one of the most notorious confidence men in the history of the West. Anyway, that is the judge's story, and what he says he sticks by. Part of the list of guests is still available for confirmation.

Life in Butte was high, wide and occasionally handsome. The days of a social life in which F. Augustus Heinze's parties, Mrs. M. J. Connell's parties, the Noyes girls' and Largey boys' parties would rival anything New York could offer were still to come, but they were definitely on their way.

And incidentally, with faro and roulette tables placed beside plate-glass windows so that passers-by could see and succumb to the lure, red-light district running openly, such dance-halls as the Clipper Shades and White Elephant debauching youth, and the screened boxes of the Casino and Comique catering to the worst instincts of the community, moral fiber was breaking down. Only on that major premise can one account for the state-wide toleration of political bribery and corruption which was to follow. Only strong character could withstand the demoralizing temptations of Butte and maintain its strength and decency.

Weakness and ignorance were in the majority, and the majority voted as advised. The tale is told and vouched for by half a dozen men still living in Montana that in one election three hundred men marched to one polling place wearing over their shoulders sashes printed with the words: "I can neither read nor write the English language. Give me a straight Democratic ticket."

Of the weak characters some were sufficiently picturesque to win a place in the history of Butte. For instance, there was "Buckets." Many old residents of the town will be surprised to know that he had another name—James Rutledge. He was a familiar figure in the streets and barrooms. Occasionally in later days he was employed by Charles and Will Clark, Jr. A popular tale concerning "Buckets" recounts his meeting one day with William A. Clark.

"Good morning, Buckets."

"Mornin', Mr. Clark."

"Have you had your breakfast?"

"No sir; not a drop."

"Hmm; are you drinking anything?"

"Yes sir; anything."

Not all reached the level of "Buckets." Some died first. The weakness of some others was revealed in different manner. "Swede Sam" Wallin was a successful gambler in Butte, with one widely known ambition—to accumulate a stake sufficient to finance a journey in affluence to his native Sweden. Time and again he won the money, shook the soot of the town from his feet, and set forth amid the cheers and congratulations of his friends to "knock 'em dead" in Sweden. And time and again he yielded to temptation for a stopover in St. Paul or Chicago, dropped his roll, and came sheepishly back to Butte.

Other gamblers won even less enviable reputations and rewards. Bob Ledigham, broke, unable to beg or borrow a stake, and unwilling to work for it, hovered upon the edge of a faro layout, watching eagerly. The deck was half out of the box when two chips that had been resting on a card, with no indication of ownership since early in the deal, seized and held his attention. He studied the case and decided that the original bet had been won and either allowed to remain for a second and double bet, or forgotten by the owner, who might have been playing half a dozen cards at the time. Such chips are known to the profession as "sleepers."

Ledigham glanced furtively at the players, decided to take a chance and picked up the chips while the dealer was settling another bet. A moment later "Spots" Sullivan, sitting beside the case-keeper, noted that the chips were gone. He looked at the other players and at the men standing behind him, some of whom occasionally were putting a bet upon a card. Ledigham stood at his shoulder. He had no great respect for Ledigham anyway, and promptly accused him of the theft. Ledigham blustered and replied in kind. Words waxed violent.

"Spots" cashed his chips and invited Ledigham to battle. Ledigham preferred verbal abuse. "Spots" left the saloon, went straight to Rich's pawnshop, bought two identical revolvers, and returned to toss one of the guns at Ledigham's feet. "That's your gun anytime you want it," he said.

A few hours later, in the darkness of a night made doubly dark by a low hanging pall of sulphur and coal smoke, young W. I. Lippincott stepped over the tongue of a wagon left at the curb of the plank sidewalk, and trod upon the body of a man. Drunks in the gutters of Butte were common, but a shiver went down Lippincott's spine at this contact. He called for help, dragged the body to the light and found no paralyzed drunk, but a corpse. It was the body of Bob Ledigham. Butte was that way, or any way its residents chose to take it, or make it.

Death came not always to Butte in such sordid guise. There were heroes among its lowly men. The tragedy of the great fire in the Anaconda and St. Lawrence Mines in the autumn of 1889 revealed such heroes. No history of any great mining camp seems complete without its story of a tragic fire in the depths of the earth. They appear as inevitable as death itself. Always they reveal a hero, or numerous heroes who give their lives with good cheer to save the lives of fellow workmen. Butte was no exception.

Early on an autumn morning Neil McCaig, a miner working on the five-hundred-foot level of the St. Lawrence Mine, adjoining the Anaconda and owned by the Anaconda company, noted the odor of burning wood in the crosscut where he was occupied. There is nothing more terrifying to experienced miners, buried deep in the labyrinthine maze of timbered crosscuts, drifts, winzes and raises which make up a great mine.

At the first sniff of smoke McCaig dropped his tools and hurried to investigate. He was alone in the crosscut, and, so far as he knew, alone on that level. Fortunately for Butte and the families of scores of miners, the night shift had gone off duty, and the day shift had not yet entered the mine. McCaig could have rushed to the shaft, signaled the cage, and been hoisted swiftly to safety. But he preferred to make an effort to save the property of the great company which paid him three and one-half dollars a day.

He found the fire in the timbers of the crosscut. He pulled off his coat and beat at the flames. Smoke rolled back upon him. Sulphurous gas from the heated rock choked his lungs.

The blaze gained headway. He staggered back to the shaft to summon help and found several men waiting for the cage, unaware of the fire. With their help he returned to the attack, connected a reel of hose to a standpipe, and turned the stream upon the blaze. Heated rocks dissolved the water into steam. The fire continued. The smoke increased. Gradually it drove the group back toward the shaft and thence to the surface.

The draft was sweeping down the St. Lawrence shaft, through the crosscuts into the Anaconda and up the Anaconda shaft, carrying fire and smoke with it. The men came to the surface with the first clouds of smoke from the Anaconda. No further alarm was needed, but the shaft-house whistles sounded a call to the town.

Day-shift miners left their breakfasts unfinished to climb the hill at a run. The wives and children of the night-shift men who had not yet arrived at their homes crowded up the slope. The word was passed that nearly all the men were out, but in the confusion of hundreds milling around the smoking shafts, few were certain that their own loved ones were safe. Then a report that there were hundred of pounds of blasting powder stored on the five-hundred-foot level where the fire was raging sent new alarm through the crowd and into the awakening city below. Explosion of such a quantity of powder would wreck the mine, impair the foundations of the surface buildings, and cut the workers off from their jobs for weeks or months. The superintendent called for volunteers to help remove the powder.

Tim Kelleher, Jerry Sullivan, John Lyons, Henry Page and Matt Sullivan, all good Irish miners, answered the call, stepped aboard the cage and dropped into the smoke-filled shaft. On the five-hundred-foot level they fought their way through heat, gas and smoke to the magazine, and carried cases of the explosive back to the cage. Kelleher and Matt Sullivan went up with the first load, leaving Jerry Sullivan, Page and Lyons to load the second cage.

At the top they awaited a signal from below to hoist. It failed to come. More smoke rolled from the shaft. Kelleher offered to go again into the inferno to rescue his friends. Matt Sullivan leaped to his side. Again the cage dropped into the smoke. On the five-hundred-foot level they found their three comrades and two others, Murphy and Shea, who had found their way from other corridors of the burning mine to the shaft. The five men were but half-conscious, near suffocation. The rescuers were in little better condition, but they managed to drag Shea and Lyons upon the platform and give the signal.

The cage lifted again in a cloud of smoke and gas. Kelleher staggered. "For God's sake, hold on," Sullivan rasped from withered lungs. "We're near the top."

But even as the cage entered the dim light of day, Kelleher fell. His head sagged over the edge of the platform. It was caught by a protruding timber, crushed to a smear of blood and brains which spurted upon the floor of the cage. The headless body dragged over the edge and plunged one thousand feet to the bottom of the shaft.

Mike Carroll, superintendent, refused to allow other volunteers to enter the burning depths, but climbed aboard the cage himself and ordered it dropped to the five-hundred-foot level. There the smoke and gas hid everything from him except the fact that the timbers were blazing fiercely, but had not yet collapsed. He himself dropped, half conscious, upon the floor, and the cage was jerked to the surface. When he was restored to consciousness he gave orders that all the openings from the St.Lawrence into the Anaconda should be closed down to the level of the fire to cut off the draft, and that the Anaconda shaft, through which the heat, gas and smoke were rising as through a blast furnace, should be sealed.

The dangerous job had hardly been accomplished when the accumulating gas and heat blew out the bulkheads with one terrific explosion. The fire spread. Four men were dead, and scores gassed and burned. The damage to the mines reached hundreds of thousands of dollars. And Butte returned to its labors and its gaieties.

CHAPTER VII

POLITICS WARMS UP

MARCUS DALY, still finding time and inclination to sit upon the curb of a plank sidewalk, borrow a chew of tobacco and swap lies with one of his imported Irish miners, was also finding time to build his great smelter at Anaconda, and thereby to intensify his feud with William A. Clark.

Clark, working day and night in the personal direction of his bank, his mines, his smelting interests and other promotions, consistently refrained from soiling his immaculate trousers upon any such lowly seat. In rare moments of relaxation he might entertain in his home, join an acquaintance for a convivial glass at one of the town's numerous bars, or even sit in at the poker game in the Silver Bow Club which was gaining some national fame. There he proved himself not a very good poker player but a cheerful loser. As his wealth, power and ambition mounted, his tastes grew more fastidious.

At the same time his interest in politics was mounting. He had thrilled to his first experience of applause when he had appeared as Montana's official orator at the Centennial Exposition in Philadelphia. He had thrilled when the newspapers gave him recognition and praise for his part in the Nez Percé war. Political ambition had been stimulated by his election to the chairmanship of the territory's first constitutional convention at Helena in 1884.

Clark was susceptible to flattery, but keen enough to discriminate and occasionally to win a loyal follower by recognizing the value of a spirit as independent as his own. Stepping late one night from the door of his bank in Butte, where he had been working for hours after the last clerk was gone, he bumped into Charley Eybel, a chef in the town's leading hotel.

Clark, relaxing after a sixteen-hour-day of labor, was moved to speak pleasantly to the late wayfarer. "You're out late, young man."

"How about yourself, Mr. Clark?" Eybel, unawed by the condescension, returned the greeting in kind.

"You know who I am? Then you might guess I've been working."

"Yes, and I've been doing something better."

"There are few things better. What have you been doing?"

"I've been courting my girl."

Clark, a keen judge, recognized a free soul. "Well, that's a good thing too. Now I'm tired and hungry. Would you join me in a bite to eat and"—recognizing Eybel's German accent—"a glass of beer. You seem to be a free soul. We'll talk it over—your system against mine."

Eybel was always willing to talk. He still is. Perhaps he was a little curious about this man who had been described to him as the richest, the hardest working and the most unapproachable in Butte. He accepted the invitation. Seated at a table in a convenient restaurant, with beer and sandwiches between them, they chatted.

Eybel told of running away from his home in Germany and working his way to America as a child. He told of experiences as a horse-wrangler and round-up cook in the cow camps of the Southwest, of riding a hundred miles with a group of cowboys to see a hanging and of staying to watch a so-called theatrical performance and the shooting up of the show. He told of guiding

English dudes on hunting trips through the Yellowstone country when it was a hunter's paradise.

Clark countered with tales of his own pioneer experiences, of carrying the mails through Indian country to Walla Walla, of bull-team freighting when the cattle froze to death under their yokes. By the time the beer and sandwiches were gone each had disclosed something of his philosophy.

"Now, Charley," said Clark at last, "I like your style. I like your background. I like your ideas, though I think you could do well with a little more sticking to the job. I like the way you look me in the eye. What can I do for you?"

"Not a thing in the world."

"What? You mean to say you don't want a better job, or some tip on the mines?"

"Not a thing from you, Mr. Clark. I like you too, but I'm too fond of my independence. I'll paddle my own canoe."

"Well, you are the first man I've met in years who didn't want something from me. I hope we'll meet often. Shake."

Yes, William A. Clark could be a mixer when he found the correct ingredients, though he never soiled the seat of his trousers with the dust of a plank curb. Charley Eybel, now in his seventies, owning and managing a little hotel in Helena, is one of some few who will attest it. For years he continued on the friendliest terms with Clark, asking nothing—and getting nothing.

In the meantime Butte continued to grow, despite the burning of the first great smelter at Anaconda almost as soon as it was completed. From 1881 to 1886 the hill's output of base and precious metals grew from $1,247,600 to $13,246,500, of which nearly half was copper. A second and larger smelter was built at Anaconda to refine the vast tonnage of copper ore coming from the Anaconda, the St. Lawrence and the Neversweat Mines, controlled by Daly and his associates.

When the new smelter came into operation production leaped. In another three years the output was doubled. A score of mines were paying dividends. Millions of dollars annually were going back into development work, into improvement of smelter operations, into the building of a greater and more modern city.

Silver had ceased to be of primary importance. Copper was king. Years of exploration within the hill had convinced the mining capitalists of the world that Marcus Daly had been right when he proclaimed it "the richest hill on earth."

Grover Cleveland completed his first term as president of the United States. The territory of Montana was admitted to the Union as a state. A third railroad, the Montana Central, built into Butte. That was the year in which Oklahoma was opened to the land rush, made familiar to a younger generation through a moving picture entitled *Cimarron.* It might be fixed in the memory of an older generation by the fact that it was the year in which John L. Sullivan, "the Boston Strong Boy," whipped Jake Kilrain, "the Baltimore Boy," for twenty thousand dollars and the world's heavyweight championship in an epic battle at Richbourg, Mississippi. It was the year of the Doctor Cronin murders in Chicago which horrified the nation as no crimes of violence had before.

More important to Butte, though its importance was unsuspected, it was the year of the arrival of F. Augustus Heinze, a twenty-year-old engineer from the Columbia School of Mines. Heinze came quietly, unassumingly, to Butte. He was merely a personable young man looking for a job. The fact that he is still described by nearly all who knew him personally, especially by the women, as "tall, handsome, distinguished, powerful," perhaps may be taken as a tribute to his personality rather than an accurate reflection of his physical appearance. His photographs would indicate that he was rather a round-faced German boy, of sturdy athletic build, and but slightly over average height. Perhaps that represented the standard of masculine good looks in "the gay 'nineties." Perhaps the photographs do not do him justice.

In any event, he was of a type distinct from the two great men of the rising city of Butte—William A. Clark and Marcus Daly. They were to become his bitter enemies, although never as bitter as they were against each other. Physically, Heinze was superior in appearance to either. Mentally, he possessed some of the more effective qualities of both. He had a finer cultural background than either. When dressed for a social function he had all of Clark's immaculate appearance and ingratiating manner. When mixing with the miners below ground, with the mill men in the refining plants, he had all of Daly's unassuming charm and popularity. In formal festivities, especially in his early days in Butte, he was a trifle shy. The women adored him, but he was a man's man.

Coming, unknown, into the mining camp he obtained a job with the Boston and Montana Company as a surveyor under ground, at a wage of five dollars a day. He lived within that income, boarding in Meaderville, an independent suburb of Butte, a couple of miles from the business center and closer to his work than the city. He was an industrious youth, with ambition equal to that of Clark and Daly, and an eye to his own future as keen as theirs.

With that eye, trained in metallurgy and mining engineering at Columbia, he was quick to see the latent riches of ore below, above and on either side of the workings in which he was employed. With that ambition he was shrewd enough to make notes of the location and direction of veins and ore bodies for his own information while he was turning in accurate maps of the underground workings to his employers. Heinze was getting his feet not only on the ground but in the ground.

Five dollars a day could not satisfy him long. He was eager to test his knowledge of the geological formation of Butte, and to raise himself from the wage-earning class. With a man named Burton he obtained a lease on a small claim in the flat. The two young partners worked earnestly and skillfully, wielding singlejack, drill and shovel, but the cost of development work soon exhausted their cash resources. They ran up a bill of two or three hundred dollars at Billy Jack's hardware store for tools and powder.

Jack, examining his books one day, noted the account and asked a clerk about it. The clerk explained that Heinze and Burton were industrious young men who stood an excellent chance of making good in the fierce competition of Butte mining. They had just opened a small but promising ore body.

The hardware merchant immediately ordered a man to go down to the lease, look over the dump and report its value. The report came back that there was selected ore on the platform worth from five hundred to one thousand dollars, much more than sufficient to meet the overdue bill. Jack ordered an attachment levied on the ore. A deputy sheriff served the papers, posted a notice on the ore platform and informed Heinze that he would be back in half an hour to remove the ore to a smelter.

No sooner was the officer out of sight than Heinze called his partner from the shaft. "We'll swap this ore for the waste dump there below it," he said, "Sweat, you miner! Shovel! Faster!"

Together the two men made the ore fly, and moved the waste from dump to platform. When the officer returned with an ore wagon, he failed to note the change. He freighted the rock to a smelter, and Heinze moved the ore to another plant. The waste was treated at the expense of Billy Jack, and instead of cash he received a bill. Heinze and Burton collected several hundred dollars for their ore. When they were ready—good and ready—they paid their bill to Jack. That was Heinze in his youth. He was not to be coerced.

In the meantime the politics of Montana were assuming importance to the leading capitalists. Taxes were going up. Assessments in Silver Bow County more than doubled in the first six years of its life, reaching more than eight million dollars in 1887, and nearly doubling again by the time the territory had become a state in 1889. And the bulk of these assessments were levied upon the various properties of William A. Clark, the Marcus Daly group, the Boston holdings and the railroads. It seemed high time that capital should take a hand in government and save itself some money.

Clark, especially, was not only ready but eager. Wealth he had now, in the millions, but his shrewd judgment demanded protection of that wealth, and his vanity demanded public recognition and praise of his accomplishments. He was still smarting under the defeat suffered in his first campaign for office, when he ran on the Democratic ticket for territorial representative in Congress in 1888, the year before Montana became a state. The fact that his own newspaper, the *Butte Miner*, attributed that defeat to "treachery in the party," was little consolation. He wanted victory, not explanation.

In the following year, when Montana had been admitted into the Union, he was elected president of the constitutional convention, and played an important part in writing that document. Politics grew heated and unconfined. Although Montana had become a state, the advance of its civilization was nothing to boast about. When its first legislative assembly convened in 1890, partisanship had attained such heat that Democrats and Republicans refused to sit in the same hall. Each party declared its representatives the only true and lawful legislature.

In those days, long before the direct primary and popular election of senators, the legislature of each state named its choice for the United States Senate. Montana improved upon that system for the moment by seating two legislatures and naming four instead of the conventional two senators. The Democrats, voting without recognition of their Republican rivals, elected William A. Clark and Martin Maginnis. The Republicans elected Wilbur F. Sanders, of Vigilante fame, and T. C. Power.

A Republican Senate in the administration of President Harrison promptly seated Sanders and Power. Clark's political ambitions had been given another slap in the whiskers. Marcus Daly, never politically ambitious for himself, but always shrewdly at work behind the lines, chuckled with satisfaction. The *Anaconda Standard*, financed by Daly and edited by J. H. Durstine, thumbed its nose derisively at the *Butte Miner.*

Daly had developed a voice for his own propaganda to compete with the voice of his enemy. As soon as the Anaconda smelter had been completed, the town laid out and a population of several hundred assembled, Daly had recognized the need of a newspaper. His projects were getting too many hard words from the opposition in Butte. But, wisely as usual, he realized that he needed an able man to edit a paper which would be a credit to his town and a power in the state. Well informed friends advised him that J. H. Durstine of the *Syracuse Standard* was the man he needed. He induced Durstine to come to Anaconda, and put the proposition before him.

"Why, no," said Durstine. "There are not enough people in this town to support a newspaper, let alone make it pay."

"There will be more," said Daly. "In the meantime we'll make good the losses."

Durstine shook his head. "You don't know how much a losing newspaper can lose. It's more expensive than a steam yacht. Besides, I don't want to be 'kept.' I want to run a newspaper that will earn its way."

Daly argued. Durstine refused all offers. Daly tried another tack. "You look around for a few days. Go over to Butte. That's a rich field. There's no reason why an Anaconda paper shouldn't circulate there in direct competition with the *Miner* and Lee Mantle's *Inter-Mountain* and the others."

Durstine's sense of courtesy moved him to agree to that. As Daly's guest he inspected the Butte mines and business field, and made himself superficially familiar with the resources of western Montana. The youth and vigor of the region and its people appealed to his imagination, but did not convince him that a newspaper in a little town like Anaconda could earn its way and maintain the self-respect of its editor. He returned to Anaconda with his mind made up to refuse. While waiting for Daly in the hotel lobby he read an item in one of the Butte newspapers that Daly had just paid four thousand dollars for a colt from a famous Kentucky stable. When Daly greeted him cordially a moment later, Durstine's viewpoint had changed abruptly.

"Is it true that you have just paid four thousand dollars for an unknown colt?" he asked.

"Yes. I'm fond of horses. I'm going to have a stable that will beat the best."

"Well," said Durstine, "if you are that way; if you will sink four thousand dollars in a colt that may or may not grow into a racehorse, I guess you can put forty thousand into an outfit that may or may not become a newspaper. Anyway, that would be worth while, if it worked. I'll take the job."

The *Anaconda Standard* was launched, and under Durstine's able editorship was soon thriving. It gave Marcus Daly a voice in public affairs in general, and against William A. Clark's *Miner* in particular. Opposing newspaper line-ups which were to become famous were in process of formation.

Montana offered a great opportunity to newspaper men. To the honest, ethical and conscientious the forthcoming battle gave a chance for public service such as many editors have prayed for, and few have found. To the venal and conscienceless it opened the pockets of men of limitless wealth.

The politics of Montana centered in Silver Bow County, although Helena, in Lewis and Clark County, was still the temporary capital of the state, as it had been the territorial capital. Silver Bow was by far the richest and most populous county in the state. There, were the greatest rewards; there, the leading powers. It was a veritable hotbed of politics. And such politics!

The majority of registered voters in Silver Bow were Irish, and Southerners who had flocked into the region during reconstruction days. They were nominally Democrats, but actually opposed in every phase of thought and character, willing and eager to fight between themselves on any local issue. The "Cousin Jacks" who had been imported direct from Cornwall and indirectly from the Comstock and California mines were nominally Republicans.

John Gillie relates an incident amusingly revealing of the character and dialect of these "Cousin Jacks," employed by the Boston interests. It happened when the Grayrock claim, worked by the Butte and Boston, was closed down in the course of a dispute with the Anaconda. Standing by as the miners left the shaft after being told that the mine had been closed indefinitely, Gillie overheard this comment:

"Good-by, birdie. Thee was a good home for we, but the savages have got thee."

These hard-working, efficient, but entirely un-American miners then were Republicans. William A. Clark, the most politically ambitious man in Montana as well as the richest, was a Democrat. Marcus Daly was an opportunist, using his personal popularity and power to swing a disaffected vote from either or both parties to his own purpose. The Republicans in Silver Bow lacked effective leadership, but were fairly well organized although in the minority in other sections of the state. In such circumstances anything might happen in politics, and almost everything did.

Local politics could make an issue of anything from a dog fight to the question of whether gambling should be licensed. The southern Democrats would fight the Irish Democrats to a standstill, with the Republicans joining in as fancy or their bosses dictated. Elections were bitterly contested and seldom decided until the last vote was counted. Not always then. In illustration is the story of the election of Milo French to the police court bench in the 'eighties.

Butte was agog in that election as it was in every election. When a naturalized Chinese known as Jimmy July came to the polls in the second ward to vote he was refused a ballot. He displayed his naturalization papers. Still some of the judges hesitated. W. I. Lippincott rallied to the defense of the Chinese. "He's an American citizen because he wanted to be, and took the necessary trouble. That's more than a lot of these bums who are voting today can say." Jimmy received his ballot and cast his vote. When the votes were counted at the end of the day Milo French was elected by a majority of one. Later Jimmy July told Lippincott that he had voted for French. With the Irish and the Missourians embattled, a lone Chinese had decided the issue. Butte was able to laugh about it and go on to the next contest. State politics were another matter.

William A. Clark fancied himself as a leader of the Democrats. This idea was unsuppressed by the refusal of a Republican Senate to accept his certification of election by the Democratic side of the legislature in 1890. Two years later he headed a Montana delegation to the Democratic National Convention in Chicago. There he helped to nominate Grover Cleveland for his second election to the presidency.

With that feather in his hat, Clark turned to the new state issue of designation of a permanent capital. And there started a battle that was a battle, the first knock-down-and-drag-out engagement of a war of political scandal and recrimination which still reeks in the political pot of Montana. Incidentally it became an issue which perhaps did more to develop the art of political cartooning in the American press than anything in our history with the possible exception of the Tweed ring scandals in New York.

Editorial license was unconfined, and unrefined. A dozen newspapers, led by Clark's *Butte Miner* on one side and Daly's *Anaconda Standard* on the other, called one another, Clark, Daly, Helena and Anaconda everything they could lay their ink to, utterly regardless of truth, decency or good manners. Twenty-one years after the issue was decided, Professor H. A. Trexler, writing in a Montana newspaper, stated that after Helena and Anaconda had won in the preliminaries, "Clark and Daly rolled up their sleeves, spat on their hands and prepared to throw mud aplenty at each other. The whole struggle is one of the most wretched in the annals of Montana. Sordid were the details and frankly selfish were the arguments."

CHAPTER VIII

THE CAPITAL FIGHT

WILLIAM A. CLARK and Marcus Daly were constitutional opposites. They were no more able to get along together than a fox and a lynx. Each had arisen through his own efforts to a position of wealth and power. Even the state of Montana, third largest in the Union, was not large enough to contain the two in peaceful relationship.

Clark's temperament required that every detail of his widespread and diversified business must be under his own hand. To keep it under his own hand he was willing to work sixteen and eighteen hours a day. He delegated no authority in business. That was a fundamental trait of his character. It revealed itself in his politics as well as his business. He wanted to see himself elected to office. He wanted to cast his own vote on important issues, not to instruct some puppet how to vote. He paid his employees and expected them to deliver the service paid for. He recognized no further obligations.

Daly had no desire for the limelight. He did desire, most emphatically, to see that no person was elected to a position of importance who did not owe that election to his influence, and who would not remember that obligation when an issue of importance to the Daly interests came up for decision. He was an appreciative and a generous man. In days before his rise to power and wealth, he had won the friendship of a man named Quinn in Virginia City, Nevada. Quinn once provided him with a grubstake. Years later, hearing from a mutual acquaintance that Quinn was dead and his family in straitened circumstances, Daly investigated, established the truth of the report, and immediately provided for the family for life. Lest they be embarrassed by any implication of charity, he advised them that it was an obligation properly owing to them because of the grubstake arrangement with the father.

In the early 'nineties, when he was planning a smelter at Carrol, he brought an engineer from the Union Iron Works in San Francisco on a contract at high salary to work with his superintendent on the job. The engineer arrived with his wife and two small children, and started work. He got along with the superintendent about as well as Daly got along with Clark. After a few months of that, Daly called him aside.

"Mr. Jones," he said, and incidentally Jones is the correct name, "you are not getting proper results with the boss of this job. I think I understand your differences. I couldn't get along with him either in your position. But I still think he is a capable man. Now I've brought you and your family way up here in the mountains and I'm going to make you a proposition in which you will be happier, and I'll be better satisfied. I've found a job for you in Butte. They can't pay what I pay you here, but if you'll go down there you will be more contented. Eventually you can work up to your present salary and more. In the meantime, I'll make up the difference in salary every month." Jones accepted, and forty years later his wife of that day still considered Marcus Daly one of the finest men who ever lived.

Daly was never in the foreground. As the smelter and town of Anaconda grew and prospered, he was seen less and less frequently. The story is told that on one occasion when he arrived at the Montana Hotel in Anaconda, the clerk, who had never seen him, answered his request for a room with the information that every room in the house was taken. "Sorry, mister, but we can't do a thing for you."

It happened that one Lou Lammers, a new printer employed on the *Standard,* overheard the remark. Lammers had arrived in the town only a few days earlier, and had taken a room at the hotel when he was unable to find cheaper accommodations in the booming town.

"I'm on the night shift, mister," he said to Daly, still standing at the desk. "You can use my room if you want to."

"Thanks," said Daly. "I'll be out and away before nine in the morning. I appreciate this very much, and before I leave I'll fix it up with the manager."

"Don't mention it," said Lammers. "It isn't anything. I rent my room by the month."

Daly occupied the room that night and checked out early the next morning. When the printer came down to the office after his day's sleep, George Reynolds, manager of the hotel, was in the lobby.

"Lou, you're pretty lucky," he said. "How so?"

"Marcus Daly left this morning, and he gave orders that your room rent was to be charged to him as long as you lived in the hotel."

"For the love of Pete!" said Lammers. "Was that Marcus Daly? I sized him up for a whisky drummer and thought he might leave me some of his samples."

Such are the tales of Marcus Daly still current in Montana. Unquestionably he was a man who commanded loyalty. He had a far-reaching vision. He was fit for the fight with William A. Clark, who also had a loyal following, and at the moment controlled material resources even greater than Daly's, and far more diversified. Clark was a mining magnate with personal control of some of the richest mines in Butte, a banker with interests extending throughout the state, great lumbering interests around Plains and St. Regis on the slopes of the Bitter Root Mountains, smelting interests in the Butte area and Colorado, and control of the vast resources of the newly developed United Verde copper deposits in Arizona. He owned the local streetcar system of Butte, and was reaching out toward the tremendous water-power resources of Montana. He was becoming a figure in the economic affairs of the nation. It seemed logical and advisable that he should control the state.

An opportunity offered in the question of location of the permanent capital. Clark, realizing that he was not very popular in Butte, came out for Helena. Daly, consistently favoring everything which might promote his town of Anaconda, announced its candidacy. The constitutional convention of 1892, of which Clark was chairman, wrestled with the issue for three weeks, and finally decided that the capital should remain temporarily at Helena, while a popular vote was arranged with all the ambitious towns upon the ballot. The two leaders in this primary were to fight it out in another state election set for 1894.

Helena led the field with 14,010 votes. Anaconda was second with 10,183. Five others trailed: Butte, 7,752; Bozeman, 7,682; Great Falls, 5,042; Deer Lodge, 983; Boulder, 295. Conclusive evidence of the heat of the campaign and the intense popular interest is revealed by the fact that 45,967 votes were cast out of a total state registration of 51,500. Probably that is a record in any state election in the history of this country. But it had served only two purposes, to nominate Helena and Anaconda as the final contenders, and to raise politics and personalities above the boiling point.

In the meantime the state's second senatorial election came before the legislature. Wilbur F. Sanders, pioneer, Vigilante, and incumbent senator who had been seated by a Republican Senate over Clark's claims to election by the Democrats in the state's first legislature, led on the first ballot of 1893 with thirty-three votes. Clark, again a candidate, polled fifteen. Several other candidates scored sufficiently to block a majority choice.

For forty weary days and nights the legislature battled, while the newspapers raged. Sanders' support disintegrated. Clark's vote climbed slowly to twenty-seven. Lee Mantle, owner and manager of Butte's *Inter-Mountain,* and one of the most influential residents of the smoke-ridden city, came into second place with twenty-two votes. There was no majority, and the strain of the battle had brought nerves and editorial pens to the breaking point.

On February 12, 1893, the *Anaconda Standard* abandoned all restraint and announced that "Clark will be elected if money can buy his election, but in no other way."

Three weeks were left of the constitutionally limited period in which the legislature could act. The deadlock continued. The legislature adjourned without an election. The *Standard* shouted with joy:

THEY FELL DOWN

Money Couldn't Buy Enough Votes
Three Votes Short
Clark Sat in Front Ready to Accept

"Helena, March 2.—The grandest fight that was ever fought in the history of Montana took place at noon today. . . .

It was a death struggle between corruption and honesty for the honor of the state. And from the terrible crisis, the greatest crisis in its history, Montana emerged triumphant. With all the forces of corruption doing their utmost for weeks, the legislature in joint assembly today declared by a vote of 37 to 32 that the majority of its legislature is honest and that a seat in the United States Senate cannot be bought. . . ."

The vote of thirty-seven to thirty-two was a vote of thirty-seven against Clark but divided among Mantle, Dixon and Carter. Clark polled thirty-two. Thirty-five was necessary to election. After fifty days of balloting the legislature had left its state without a senator. Governor Richards promptly appointed Lee Mantle, who had polled twenty-five votes on the final ballot. He succeeded Wilbur F. Sanders. Clark's *Butte Miner* loudly proclaimed a Daly plot.

Mantle was a man of parts, handsome, powerful, shrewd, democratic, popular. As a youth he had served efficiently as stage agent and telegraph operator in Pleasant Valley, Idaho.

When the stage station had closed with the development of the railroads, Wells-Fargo's traveling superintendent had introduced him to the opportunities of Butte. He had organized the *Inter-Mountain* as the first Republican newspaper there, and had built it into a profitable organ. Incidentally he acquired valuable mining properties. Although a bachelor, he had attained social as well as political distinction. Mantle knew not only the beans which constituted an important part of frontier diet, but the delicacies which he might encounter in the grill of the Senate. Witness a souvenir menu of December 12, 1891:

STAG PARTY
Tendered by Lee Mantle and W. McC. White
On the occasion of the Anniversary of their
advent on earth, being their (?) birthday.

***

Sauterne — Blue Points on Shell
Celery

***

Consomme, à la Royal in Cup

***

Riesling — Boiled Red Snapper
Hollandaise Potatoes

***

Frog Saddles, à la Poulette
Hericot Verts

***

Pontet-Canet — Vol au Vent of Sweetbreads
Petit Pois

***

Roman Punch

***

Broiled Quail on Toast

***

Burgundy — Filet of Beef Larded, aux Champignons

***

Turkey with Cranberry Sauce

***

Chicken Salad, Mayonnaise Dressing

***

Rum Omelette

***

Dessert
Vanilla Ice Cream — Assorted Cakes

***

Pomery Sec — Fruits, Nuts, Raisins, Candy
Roquefort — Crackers
Spanish Olives — Cafe Noir

***

Courvoisier

Evidently both the hosts and the chefs did their best, and it was an achievement worthy of the effort. The fact that they mixed the French and English in their printed souvenir menus detracted not in the least from the gay success of the party. Perhaps it adds to the value of the item as a social document indicating that Butte was coming into the big money and beginning to feel its oats.

From Harry Edgar's notable meal in Bannack in 1863—Salt Lake eggs, ham, potatoes and everything! Such a supper!—to a thirteen-course banquet, with the correct wine for each course, in Butte in 1891, was quite a step for mining camps. Butte had accomplished it. Butte still had far to go, but was very much pleased with itself as it was.

Even the panic of 1893 did not greatly disturb Butte. Its silver production had been greatly reduced but it had something in inexhaustible quantities that the United States needed—copper. Evidence of that need may be cited in the fact that though electric power and telegraphic service had been greatly extended in recent years there was still no such service to the Hawaiian Islands, then still referred to as the Sandwich Islands. News of the revolution in Hawaii and the overthrow of Queen Liliuokalani, preparing the way for American annexation, was brought to San Francisco in that year by steamer. So peaceful did the Pacific seem that the United States maintained but one modern warship in that ocean, the cruiser *Boston.*

But though there were riches in Butte, and evidence of improvement in the social amenities, there was no peace. Political morality had not kept pace with social accomplishment.

Clark, still smarting under two lashing defeats in efforts to reach the United States Senate, planned to organize the state more to his liking, and at the same time show the presumptuous Marcus Daly, Irish immigrant, that he was not so great as his actions indicated. When the final campaign for designation of a permanent capital started in 1894, Clark was prepared. But Daly also was prepared. Immediately after the preliminary election of 1892 had brought the issue down to a clearly defined contest between Helena and Anaconda, Daly's *Anaconda Standard* adopted the practice of referring to Helena, when Helena forced itself into the news, as "temporary capital."

As the campaign warmed, the Daly forces began to distribute a specially made cigar which was labeled the Anaconda-for-the-Capital Cigar. The Clark forces immediately announced that the cheroots were made by "scab" labor, thereby seeking to prejudice the laboring classes against Anaconda. The great strikes in the Coeur d'Alene mines and in the Carnegie steel works at Homestead, Pennsylvania, had made the Butte workers intensely union-conscious. They had never had a strike up to that time but they were well organized and inclined to give credit for their prosperity to their unions.

In the Homestead strike three hundred Pinkerton detectives, guarding the steel mills from the strikers, had engaged in a battle with the men in which ten were killed and sixty wounded. In 1894 occurred the bloody Pullman strike, in which Pinkerton employees also took up arms against the unions. Butte miners and both Butte and Anaconda smelter men were acutely conscious of the war of labor against capital going on in other parts of the country, and they were most bitter against the Pinkertons.

So the *Missoulian* printed a story that the Daly forces were hiring Pinkerton detectives to register voters for Anaconda illegally. The *Montana Silverite,* a Populist newspaper in Missoula, countered with the assertion that Helena, "Her Hogocracy of Last Chance Gulch," was of importance only as headquarters for Chinese laundries, gambling dens and opium joints.

So it went, hammer and tongs, throughout the state, to the hilarious delight of the fighting Irish, the sardonic smiles of the Missouri Democrats, the profit of almost everyone who could influence more than his own vote, and to the deepening embitterment of the Clark-Daly feud.

A sidelight, probably typical of what was happening in every section of the state with either Clark or Daly as its dominating influence, is narrated by Thomas Sullivan, now a retired federal forest service employee living upon his own land in Glacier Park. Tom Sullivan, who may be remembered by many of the pioneers as "The Bald-Faced Kid," a popular jockey in the informal race meets of the early days, had run away from his home in Duluth and come to grow up in Montana. He had arisen from the status of a handy boy around the stage stations before the days of the railroads, to the temporary heights of an itinerant jockey, and thence to the exalted position of a stage driver.

When the Great Northern was building west, Sullivan had driven stage from Shelby, and later from Cut Bank, to Libby, near the western boundary. During the winter months the grading for the line on the frozen plains near Shelby had been stopped, and the crews transferred to the mountain area for hard rock and timber work. Laborers were constantly drifting out to the end of the railroad. Many, finding no work there, and being told that there were jobs in the mountains to westward, continued their journey on foot. Not a few of these men, insufficiently clad, and undernourished, succumbed to the hardships of that long hike, and froze to death beside the road.

The stage company's orders to its drivers were to stop to pick up a dead man, but never to give a free ride to a live one. Stopping to pick up a dead man, according to Tom Sullivan, was not so simple as it sounds. There were six horses to each stage, all selected stock, full of oats and energy to maintain the schedule. When the stage was stopped along the road, every horse had to be hobbled to prevent a runaway. Hobbles were carried on the hames of each horse-collar to be handy. By the time the horses had been hobbled, the dead man picked up and placed securely on the stage, the horses unhobbled, and the drive resumed, considerable time had been lost. But the schedule had to be maintained. That necessitated some wild driving over the frozen rutted road. Tom Sullivan, past seventy, grins as he thinks about it.

"I remember once I had a negro as a passenger. It was mighty cold, but he wanted to ride on the box with me. 'Dey's ladies an' white men inside, boss,' he said. 'Dey don't want me inside.' Well, I could understand how he felt, and his fare was paid. 'All right,' I said, 'you can ride with me if you want to, but it's going to be rough—and cold. You'll have to be strapped to the seat.' Of course the driver was always strapped. Otherwise those horses would pull him clear off the seat when they were fresh, even if the chuckholes didn't throw him off.

"Well, this negro climbed up and fastened the straps around him. By the time we got to the next station he was the whitest black man I ever saw, or you might say the grayest. He unfastened the straps and climbed down. Then he looked up at me. 'Boss,' he said, 'I wouldn't have your job if they gave me the whole of Montana and the rest of the United States throwed in.'

"Another time, I remember, I stopped to pick up a dead man, half covered with snow alongside the road. There was a woman passenger inside, so I strapped the body on the boot. The woman discovered what was going on, and she gave me hell. 'That's no way to treat a dead man, like a log of wood,' she says. 'What if he was my husband?' Well, I'd had some trouble with the horses, and maybe I was a bit peevish. I says, 'Well, ma'am, even if he was he might be satisfied to ride outside: Then she did lay me out. 'All right,' I says finally, 'if you want him inside with you, I'll fold him up so's he can sit on the seat, and put him in.'

"That didn't set so good either. There just naturally was no pleasing that woman. She gave me the devil every chance she got, all the rest of the trip. When we came to the last station the boss says to me, 'Tom, this woman you got has to go ten miles up the north road to a ranch. You hitch up the buckboard and take her.' I'd have quit my job first, and I told him so. 'Hmm,' he says. 'All right; I'll take her.'

"The fact was that good stage drivers were hard to come by, and he knew it. He took her to where she was going, off the line. When he got back he says, 'Tom, you were right. That woman certainly gave me hell over the whole trip.'

"But they weren't all that way. I used to have a good many passengers that I hardly knew were in the stage. It was orders not to pick up any of the tramps. So when the temperature was down below zero and maybe a blizzard blowing, I'd just slow down and look 'em over to see that there wasn't a dead man amongst 'em. Then I'd move on. When I'd get near to a station I'd find four or five of 'em had climbed onto the boot. So I'd cuss 'em out good and throw 'em off, explaining my orders and telling 'em to hustle on up the road beyond the station before they froze to death. When I'd changed horses and got out of sight of the station, there they'd be again. So I'd slow down to bawl 'em out, just for luck. Funny thing, they didn't seem to take it seriously at all. When I'd get near the next station I'd look around, and there they'd be on the boot again, and the whole thing to do over. Maybe it was just as well. Probably I didn't have to stop to pick up so many dead men that way."

That's the sort of man Tom Sullivan was, and is. The sketch is given so that readers may be better able to draw their own conclusions as to the accuracy of his recollections concerning the Clark-Daly fight to establish the permanent capital of the state. By that time the railroads had put most of the stages out of business. Sullivan was logging on contract near the lumber town of Plains. He was backed by James A. McGowan, a general-store keeper in Plains, but could finance only a small operation, in which eighteen men were employed.

"At that time William A. Clark controlled extensive lumber interests around St. Regis and Plains," says Mr. Sullivan. "When the capital fight warmed up he sent whole wagon loads of liquors and cigars to the camps of the contract loggers and handed them out to the men. My crew had so much free booze and tobacco that they were no good on the job. I went to McGowan and made a kick. McGowan told me to keep quiet and go along, even if I wasn't getting out any logs worth mentioning. Well, I did it, and pretty soon I got word that there ought to be two hundred and fifty men on my job instead of eighteen.

"'Fine,' I says to McGowan, 'but who is going to pay the bills for grub, building camps, roads and so forth?' He said, 'Never mind. Go to the store with a wagon and get whatever supplies you need. You can draw on the bank for the payroll if you have to at the end of the season.'

"Well. I did it, and put on two hundred more men. The Clark interests built the roads, while we cut the logs and piled them on skids, ready to take out. All that was necessary was to keep plenty of men on the payroll. With credit for supplies, and roads being built without cost to me, I could see that I'd have plenty for the payroll and a good fat profit. When the election came around the crew was able to cast two hundred votes for Helena for the state capital. Any men who had not been on the payroll at least thirty days before election were told what names they were registered under, and voted accordingly."

And Helena won. The repeated political slaps given to Clark's ambition by the Daly interests were answered in kind, at a cost which has been estimated in the neighborhood of one million dollars to each side. But Clark could crow, and did. His *Butte Miner* of November 7, 1894, called editorially for:

"Three Cheers!

"The people are supreme!

"The Citizenship of Montana is vindicated!

"Tyranny has reached its Waterloo! . . .

"This election in Montana is not only the Waterloo of the most tyrannical corporation that ever attempted to crush out the independence of the people, but it is, the declaration of independence of one of the grandest people this world has ever seen. . . ."

Helena's celebration was worthy of its accomplishment. A bonfire which could be seen throughout the length and breadth of Prickly Pear Valley blazed upon the summit of Mt. Helena. One thousand cheering citizens greeted the special train which carried Clark from Butte for the festivities.

"When it arrived and Mr. Clark stepped to the platform," said the *Miner's* account of the celebration, "he was taken upon the shoulders of several strong men and carried through a mass of cheering people to the carriage in waiting for him. . . . The horses were taken from the carriage and a rope attached to the pole. Then, drawn by hundreds of men, the carriage proceeded in triumph through the streets of Helena to the hotel with the bands of music playing and men and women cheering the heroes of the hour."

Fireworks, a great parade, and a reception at which twelve thousand persons jammed the auditorium to hear Clark's speech of triumph concluded the festivities.

In amusing contrast to the *Miner's* story, the *Anaconda Standard* commented under an obscure headline and a Helena date line: "The ovation which W. A. Clark received early in the evening upon his arrival from Butte showed that Helena thanks Clark for sacrificing his own city for Helena. He rode in a richly decorated carriage."

There was no thought in Montana of letting bygones be bygones. This was not a matter of sportsmanship but of politics, having no connection with sportsmanship. There was no suggestion that Clark and Daly should kiss and make up. Too many lesser men had profited from the battle and hoped to see it renewed.

The breach was widened, and into that breach was climbing no less a man than F. Augustus Heinze.

CHAPTER IX

HIGH SPIRITS AND TRAGEDY

BY THE time the capital fight had been settled in favor of Helena, to the discomfiture of Marcus Daly and the great delight of William A. Clark, Butte had established itself as the greatest copper-producing center on earth. Thirty more years, with a total production of more than two thousand million dollars, would prove beyond dispute its right to the title of "the richest hill on earth," but it was already assuming the substantial qualities and airs of that title.

Butte was bumptious. The value of its silver, which had been shrinking steadily for several years, had collapsed abruptly with the panic of '93, but the value and extent of its copper had far more than made up that loss. It considered itself superior even to a national panic. Its resources were safe and sufficient. Even its cash was only temporarily restricted by that panic. Its leading men were able to overcome the difficulty.

William A. Clark's credit in New York and London was sufficient to buy four hundred thousand dollars in gold coin in London, and have it shipped in kegs to Butte. There he opened the kegs in a hotel room and cashed checks sufficient to meet the urgent needs of local business men. He did not even require that they have an account in his bank. Clark was personally and thoroughly acquainted with the credit rating of virtually every man of importance in the territory. He was taking no chances, but was stimulating business and making friends who might later be valuable in both his business and his politics.

Marcus Daly, not to be outdone, shipped in seven hundred thousand dollars in gold coin, and improved upon Clark's achievement by that much margin. Butte arose superior to the first stringency of the panic, and went on with its primary business of producing copper. Its thousands of miners and mill workers were all employed, and at fair wages for the time. Local business, especially the saloon business, was thriving.

Paid advertisements in the guise of news notes in the local papers reflect that phase of community life. "Next door to the Miners' Union building is one of the cozy places to spend an enjoyable evening in a social game of billiards or pool. A first class bar, where the miner may be refreshed with the best of whiskies, Milwaukee, St. Louis and 'home-made' beers. The proprietor is always courteous and accommodating, and it is his endeavor to make things agreeable for his customers." Or this: "James Lynch's corner is still the popular station for the boys going up or coming down the hill. The excellent quality of the goods kept in stock is a strong drawing card."

Thrifty miners were acquiring homes and accumulating savings. Others were enjoying what they looked upon as a gay life. Nearly all had money, at least on pay day. But Butte was no melting pot. The Cornishmen, for the most part, lived in one neighborhood. The Irish workers kept more or less to themselves in another section. The Missourians and other Southerners were equally clannish. Southern Europeans who were coming in in greater and greater numbers for the hard labor occupied another area. But all were feeling their oats, and perhaps their rye, bourbon or Scotch.

On the Fourth of July, 1894, they demonstrated in a riot which left three dead, many injured, and filled every cell in the local jail to suffocation. A grand time was had by all who survived, especially by the Irish firemen. Butte was that way.

The trouble has gone down in the history of Butte as the A. P. A. riot. In 1894 the American Protective Association was to Roman Catholics as a red rag to a bull. When Simon Hauswirth included in the Fourth of July decorations of his Columbia saloon on West Broadway the letters A. P. A., formed in bunting, trouble was not far away. The Sazerac saloon, conducted by Anderson and Keeton, prepared for the same thing with the same letters formed of small American flags in the transom above the door.

About mid-afternoon indignation in the form of a stick of dynamite exploded under Hauswirth's plate-glass window, shattering it into a thousand pieces. Immediately William Page, a driller, William Ferguson and others mixed in a joyous free-for-all battle, in the course of which a revolver bullet plowed a furrow in Page's scalp. Policemen arrested Ferguson, fought off the howling mob, and dragged him away to a cell.

The "Cousin Jacks" and the fighting Irish clashed in a dozen saloons and alleys. For three hours "pandemonium reigned," according to the subsequent report of the *Butte Bystander.* Sheriff Reynolds telegraphed the governor for the militia. "Clubs, rocks, fists, revolvers and knives were used continually."

Mayor Dugan, D. J. Hennessy and other citizens finally induced Hauswirth to allow the offending letters to be removed from the front of his saloon in the interest of peace. The mob then concentrated on the Sazerac, but Anderson and Keeton refused to change the offensive arrangement of their flags. Judge McHatton and Mayor Dugan pleaded with the mob to go home, assuring them that justice would be done, but the mob had other ideas about that.

By that time half the policemen in town were wearing black eyes, a dozen men had been jailed for disturbing the peace, and local members of a militia company had been ordered to assemble. Even the elements took part. A thunderstorm poured a drenching rain upon the embattled miners, with no apparent dampening of their anger or ardor.

Sticks and stones showered upon the front of the offending Sazerac. Its defenders rushed the mob. Fighting groups milled in the street. "In each fistic encounter," said the *Miner's* account, "the crowd swarmed about the fighters and those close by usually took a hand until the officers captured the belligerents. Then, as there was no chance for the prisoner to fight back, he would be beaten almost to a standstill by three or four of the crowd and taken to jail."

Amid this turmoil one Frank Morford, a baker, and a member of the militia, was ordered by a superior officer to get into the crowd and tell all his fellow militiamen whom he could find that they were ordered to assemble at once at the armory. Morford was promptly walloped on the side of the head. Before he could regain his balance a blow on the other side straightened him up. Thereupon he drew his gun, and in the excitement which followed two shots were fired, one of which creased Morford's scalp. Then, but not until then, according to his own statement as subsequently published in the *Miner,* he began to shoot in self-defense.

Officer Daly fell with a bullet through his heart. Samuel Dunn, described as an innocent bystander, fell with a bullet through his stomach. Half a dozen guns were out, and bullets were flying in every direction. Joseph Crouch dropped with a bullet in his leg. The situation had passed the point of a joyous Donnybrook Fair. The crowd began to fall back. Officers seized Morford. The crowd closed in again, everyone within reach taking a swing until the prisoner's face was beaten to a pulp, made more gory by the stream of blood from the bullet wound along his scalp.

Again the crowd rushed the Sazerac, with the evident intention of razing the entire building. An alarm brought the fire department, but instead of turning their streams upon the mob the firemen turned them upon Hauswirth's saloon, from which the offending letters had been removed, but not forgotten. Flags and bunting of the remaining decorations were torn away by the high pressure streams. Then the firemen, while the mob raged and fought in front of the Sazerac, turned their streams into Hauswirth's place. But Simon had no such sense of humor. He came forth as promptly as a rattlesnake from a flooded gopher hole, and just as mad and dangerous.

The firemen retreated and directed a devastating stream into the Sazerac. A volley of bullets answered. Captain Lyons stepped up and deliberately aimed four shots at the men inside the saloon. Every window, glass and bottle in the place had been broken, either by the stream of water or by bullets and stones.

By this time a platoon of militia, in full uniform, had been mobilized. With fixed bayonets they charged the crowd, sweeping rioters, firemen, spectators and all before them until Broadway was cleared from Main Street to Academy. Butte's high spirits, religious prejudices and sense of humor had had their fling.

Another, more spectacular, more heroic, and far more tragic view of its firemen was to be presented a few months later.

That, according to the newspaper headlines, was "Butte's Night of Horror."

A fire alarm about nine o'clock of a freezing January night in 1895 called the department to a small blaze near the warehouses on the southern edge of the city. Bystanders warned the firemen that there was blasting powder stored in one of the warehouses. But the firemen hesitated only an instant, and attacked the blaze. It spread, despite their efforts. By the time the Kenyon-Connell and the Butte Hardware Company warehouses were in flames, every piece of fire apparatus and every fireman in the city was on the job.

An hour after the first alarm, spectators at a distance, watching from their homes on the hillside or from points in the valley, saw a burst of flame, as if from a volcano, rise abruptly to a height of several hundred feet and spread out in mushroom shape. A terrific concussion came while the flame seemed to be hanging motionless. Windows through which spectators were gazing crashed in their faces. Butte shook to its foundations. Men deep in the mines, hundreds of feet below the surface, felt the concussion, and rushed to the shafts, fearing a mine disaster of tragic extent.

A second explosion, as heavy as the first, seemed to expend itself nearer the ground. Every window within a radius of half a mile, and some two or three miles away, which had not collapsed under the first concussion, went out with the second. Torn fragments of the fire department horses, arms and legs, and headless bodies of firemen and spectators rained upon Butte with the shrapnel of iron from the hardware supplies.

Tons of iron plates, four by six inches in size and half an inch thick, for use in railroad construction, had been stored next to the giant powder in the hardware warehouse. These had been spread with the devastating effect of a huge explosive shell over an area of many blocks. Some were picked up, twisted and broken, more than two miles away, on the following morning.

They had cut down the firemen, fire horses and near-by spectators, broken through the walls of neighboring buildings, and spread death, injury and destruction through the whole lower part of the city. The fire chief's hat was found later, filled with blood. His body was never found. Arms, legs and bodies, so mutilated as to be impossible of identification, were found all around the neighborhood. One spectator was knocked flat by a man's dismembered leg, and himself uninjured. Only one horse of the entire fire department equipment was left alive.

This animal remained a pet of the city as long as it lived, wandering freely through streets and alleys, and receiving handouts of apples, sugar and other tidbits at the doors of restaurant kitchens and private homes. That was Butte when it recovered from its shocking disaster.

But before daylight dawned City Marshal McNichols had organized a search of every house, damaged or undamaged, within half a mile of the explosion, and of every abandoned prospect hole or shaft in the same area. He estimated that fifty or sixty persons had been killed and hundreds injured. Fifteen, critically wounded, were in the Sisters' Hospital. Mayor Dugan had appointed a committee of relief.

Among the victims at the hospital was Thomas J. Riley, with a leg so badly mangled that it required amputation. Prominent on the committee of relief was Patrick Largey, banker and merchant. There were the seeds of another drama.

When Butte had buried its dead, patched up its wounded and repaired its wrecked buildings, there was still trouble.

Storage of explosives within the city in such quantities as had been revealed by this terrific blast was strictly against the city ordinances. Investigation was demanded. The Butte Hardware Company explained that its chief storage magazine was at some distance outside the city, and proved it. But all the smaller mines and many of the larger ones were constantly sending to the company for explosives. They could not easily send out to the main storage depot for one box of dynamite at a time. Naturally they brought in a load, and parceled it out as it was called for.

Patrick Largey was interested in the Butte Hardware Company. Patrick Largey was a member of the relief committee. He had contributed generously to the care of the injured. Most of Butte, returning to its own pressing affairs, was willing to let it go at that. Not Thomas J. Riley, with a leg sacrificed and health impaired by internal injuries. He had been provided with a small income but he had not been provided with a good leg or a forgiving disposition. When he was able to hobble around on a crutch he hobbled into Mr. Largey's bank.

"I want ten thousand dollars," he said.

"Well, so do I," said Mr. Largey, smiling a banker's smile.

"I mean it," said Riley. "You've got it. I want it. It was your dynamite that blew my leg off. I want ten thousand dollars, and want it now."

"You're being taken care of, Riley," said Largey, his smile gone. "Go on home. You can't hold me up for ten thousand dollars."

"You won't pay me?"

"No."

"Then you'll learn how it feels," said Riley, and shot the banker through the body. Largey went to his grave, and Riley, with sardonic cheerfulness, went to state's prison for life. Largey's sons remained to cut a swath in Butte society, with sterling silver favors costing fifty dollars each at dinner parties, until they reached approximately the same economic status at which their father had entered Butte. Other prominent citizens who had received threatening letters from victims of the explosion who held that their losses and injuries were due to disregard of the law by the men who had stored the powder within the city limits noted Pat Largey's death, and decided that Butte was no place for them. Several wealthy residents sold out their local interests and moved out of the state.

Butte returned to its business and its pleasures. That was the civilization of Butte in the gay 'nineties, a far step from the ox-team, beans and whisky civilization of less than twenty years earlier. It was a civilization peculiarly adapted to the ambition of F. Augustus Heinze, now definitely rising toward the position of a local Napoleon of finance and industry.

CHAPTER X

THE RISE OF F. AUGUSTUS HEINZE

HEINZE had occupied himself with energy, judgment and discretion in his first years in Butte. He had gained a comprehensive knowledge of the geological formation of the big hill and of the underground labyrinth. He was especially well acquainted with the Boston and Montana Company's holdings, where he had been employed as a surveyor. He had gained some knowledge of the difficult problems of smelting the copper ore. With the depression of silver values before the panic of '93 he decided that scientific handling of the copper ores would make countless fortunes in Butte. Therefore, he should build a smelter.

Thereupon he returned to New York to raise the needed capital, and succeeded in interesting the New York agents of Baring Brothers, the famous British bankers. That in itself was no small accomplishment for a youth of twenty-two years. The Barings were important men in the financial world, and had been for more than a century, second only to the Rothschilds as international bankers. Evidence of this importance is revealed in the fact that when a collapse of their extensive South American securities occurred, shortly after Heinze made his trip to New York, the Bank of France, the Rothschilds and numerous European joint stock banks guaranteed a fund of ninety million dollars to tide them over the crisis and prevent an international panic. In that situation naturally any tentative promise of financing Heinze's proposed smelter and campaign for seizure of "the richest hill on earth" failed to materialize.

Heinze, disappointed but undismayed, took a staff job with the *Engineering and Mining Journal* for the threefold purpose of earning a living, studying stock-market practices and finance, and making acquaintances who might eventually aid his plans to capture Butte. He joined the staff in 1891 and remained a year. An article concerning him, printed in that same *Journal* twenty-three years later, said:

"He is recalled as the only man who habitually came to the office wearing a top hat and frock coat. He was at that time a handsome young fellow, genial in manner, popular among his associates, but exhibiting no great liking for hard work, and being especially averse to anything confining."

Toward the end of his year in New York the death of a grandmother left him fifty thousand dollars, the first important money he had ever controlled. By that time he was pretty thoroughly acquainted with Wall Street practices. He had come to the realization that the most advanced scientific understanding of metallurgy then available in the world would be necessary to accomplish his purpose. With part of his legacy he journeyed to Europe, and took an intensive course of study in a German university.

With that equipment and the bulk of his legacy he returned to New York and organized the Montana Ore Purchasing Company to market Butte copper on the Atlantic Coast. In that business he was sufficiently shrewd and successful to attract the attention of Lewisohn Brothers of Boston, the leading copper sales organization of the world. But that was only an additional stimulus to the Heinze ambition.

He returned to Butte and leased the promising Estella Mine from James Murray. At the moment Jim Murray was one of the outstanding figures of Butte. He had entered the town as a gambler, made money, engaged in real-estate speculation with large profit, and taken a hand in the mining business with an eye to larger profits. His admirers even pointed him out to strangers as "the shrewdest operator in Butte."

Murray was a handsome man, admired by men, adored by women, lavish in his gifts to personal charities, sophisticated, romantic. But though his shrewdness was a byword, it failed to measure up to that of Heinze.

Murray had been drawing twenty percent royalties from a lease on the Estella. When the lease expired Heinze offered him fifty percent on all ores running fifteen percent copper, and nothing on so-called second-class ores, returning less than that percentage. As the mine was then in first-class ore, the new arrangement seemed attractive to Murray, and the papers were signed. From that time until the expiration of the lease, every ton of ore hoisted from the mine consistently ran under fifteen percent. Heinze took it all.

Heinze's miners simply caved down the country rock beside the vein and mixed it with the ore to be sent to the smelter. Murray, getting nothing from his property, investigated and discovered how he was being duped. He promptly sued. Heinze obtained the best counsel available, and won the action. It was his first lawsuit, the first of scores which were so to clutter the Montana courts that minor independent suits were sometimes delayed for years.

But Jim Murray was a gambler. This was not the first game in which he had lost. He grinned and took it, with the comment, "If I'd known that young fellow ten years ago I'd have owned all of Butte." That was precisely what Heinze himself was planning to do without the assistance of Murray or anyone else.

But he needed more money. Butte business was big time business. Clark had millions. Daly controlled millions. The Boston and Montana and the Butte and Boston companies represented more millions. Compared with them financially Heinze was a piker. Compared to them intellectually he was their equal if not their superior.

With his lease on the Estella working so profitably, Heinze took another lease on the Glengarry, which had been closed when it ran out of profitable ore. With his keen understanding of the geological formation of the hill, he soon opened a rich ore body in that property.

Smelting costs in the district were still high. Clark's Colorado and Montana Smelting Company's plant against the foot of the hill south of town, and Daly's Anaconda smelter had almost a monopoly of the business. Although these two men were bitter enemies, neither would cut off his nose to spite his face. Neither was willing to promote a price-cutting war to put the other out of business. The few small independent smelters had almost all they could do in handling the ore from the mines of their owners. Smelting costs had in fact fallen to approximately fourteen dollars a ton, which was half what Clark had charged at his first large plant.

But Heinze believed that that price was still double what it should be. It was that belief which had moved him to go to New York to raise funds for an independent customs smelter. Now, with hundreds of tons of ore coming from his Estella and Glengarry leases, he was more than ever convinced that he was handing out far too much money for their treatment. And incidentally too much copper was being lost in that treatment. An accidental discovery of a scientific fact overlooked by all the metallurgists, mill men and smelter men in the world gave proof of that. Jim Ledford, a horny-handed, clay-smeared miner, who has never been properly recognized or honored for his great contribution to the science of metallurgy, was the discoverer. It was Ledford who revealed to the world the secret of recovering copper by the leaching process, a method which added countless millions to the copper production of Butte and of the world. His reward was ninety thousand dollars, and that only because he was a close-mouthed man.

In the back yard of Ledford's cabin on the slope below the Anaconda Mine, was a dump of discarded cans and iron junk. Waste water, pumped from the mines, so impregnated with minerals, chiefly copper sulphate, that it was useless for irrigating or any other ordinary purpose, ran through this junk heap. One morning Ledford noticed that most of the junk had disappeared, and in its place was a heavy deposit of slush.

"The sulphur," one writer on the subject years later explained most vividly, "had dropped its copper playmate and run off with an iron affinity."

Ledford, curious, obtained an assay of the deposit and found that it was ninety-eight percent pure copper. He kept his secret, and made a year's contract for all the water flowing from the Anaconda Mine. In that year he cleaned up ninety thousand dollars. Having discovered what was going on under its nose the Anaconda refused to renew the contract, but Butte and the world had learned something new and profitable concerning metallurgy.

Tons of arsenic, tons of sulphur and other by-products were going to waste in the smoke of the roasting ore heaps and smelters. Improvements in the methods of smelting ores were also made to capture those stinking and poisonous millions of dollars. All these possibilities appealed to Heinze's scientific training and sense of business. Coupled with his firm conviction that he was paying twice what he ought to be paying for the treatment of his ores, these possibilities of profit stimulated him to new effort to build a smelter of his own.

By this time, with the profits of his leases proving his mining ability, and the contacts he had made while on the staff of the *Engineering and Mining Journal* indicating his financial ability, he was able to raise sufficient money to build a plant low on the slope below the Anaconda Mine. At about the same time he obtained temporary control of the Rarus Company, operating a mine between his smelter and the Anaconda.

Still he was not in the big money. He was, to be sure, somewhat of a social lion in Butte, but by no means an economic or political lion. Even as a social lion he did not roar. Perhaps some of his popularity was due to that fact. He was, if anything, a little shy. At least he appeared to be. Dressed for the evening in the most carefully tailored raiment obtainable in New York, he was handsome and charming. His travels, education, and social training in a home accustomed to the best, formed an excellent background for a young man entering the upper crust of Butte society, in which the sons and daughters of the leading citizens were taking their tone from the most expensive colleges and finishing schools of New York, Boston and Paris.

The women found his trait of shyness, coupled with his good looks and fine tailoring, especially delightful. For one thing, it gave the maids and young matrons an advantage which they were quick to seize. Heinze never seemed to know how to extract himself gracefully from either tête-à-tête or admiring feminine circle.

The girls loved that, and seldom failed to take advantage of an opportunity. Heinze was the only man under thirty years of age in Butte who was making a bid for wealth and power in his own right. William A. Clark's sons, Charles and W. A. Jr., were a rich man's sons, well educated, traveled, active, socially winning, but somehow lacking in the appeal which Heinze's independent rise made for him.

Heinze was a man's man. He was no philanderer. No social or romantic scandal was ever heard about him in Butte. Probably both maids and matrons laid their traps, but he was not to be caught. An invitation to a dinner party in his apartments was an event which might require a hasty journey or at least a telegram to some modiste in New York to obtain the latest in evening gowns, but the party itself was always circumspect, adequately chaperoned.

Chaperons were an important institution in Butte society in the gay 'nineties. In spite of the fact, or possibly because of the fact that mistresses were condoned, no miss who valued her reputation took unnecessary chances. For Heinze was not the only man rising to wealth and power in the growing city who maintained some of the comforts of a home without the responsibility of a wife. Others whose names were great and are still great in the annals of Montana, though they may be enjoying an affluent and respected old age in southern California, were permitted the same privileges. But Butte was careful. A handsome woman, handsomely gowned, might be pointed out in the better shops or in the theaters as so-and-so's mistress, but she was never seen elsewhere.

Young matrons of unquestioned social position served as hostesses and chaperons when the bachelors entertained. The mistress disappeared temporarily into complete retirement. Even her photographs were removed. To attend to the latter detail, according to Mrs. T. J. Murray, one of the handsome daughters of John Noyes, a leading pioneer of Butte, was the first duty of the chaperon. She must arrive before the other guests and see that no indication of a less refined detail of life was left in evidence. For Augustus Heinze's circumspect little dinner parties, this task became somewhat difficult as time went on.

"Fritz's last girl," says Mrs. Murray, "had either a sense of humor or a sense of resentment. She would tack up photographs of herself on the walls, so high that we had to find a stepladder to take them down. But none of these women ever caused a scandal, either for Fritz or for other important men who were known to be similarly situated. Girls who were guests have told me, and I have myself, on occasion, sensed that there was a woman in the background, but she never appeared. The dinners were always beautifully arranged, perfectly cooked, perfectly served, and the parties were gay and lovely. Fritz knew how to entertain."

Public scandal was left to the bartenders, barbers, their women and others of their social status. The Butte and Anaconda newspapers for days were filled with lurid stories when Walter H. Brooks, a bartender in the Casino saloon on East Galena Street, shot and killed Emery Chevrier, a barber, in Chevrier's room in the Maule Block on West Park Street. All the scandal that the most depraved taste could assimilate was dished up in that affair. A Mrs. O'Reilly who also had an interest in Chevrier, had shadowed Mrs. Brooks to the barber's room and led the bartender there. While his wife screamed and begged, Brooks killed the philanderer. The story trailed on for days. A coroner's jury held Mrs. O'Reilly as an accessory. A Mrs. Johnson also was arrested. Butte must have its morbid thrills, although it could not get them through the mistresses of some of its leading men.

Social activities were merely an incident in Heinze's crowded days and nights. He had convinced himself even before he convinced Butte that he was the man of the hour. With his Butte smelter operating, to the greater profit of his leases in the Estella and Glengarry, he extended development of the Rarus Mine, up the slope above the smelter, toward Daly's Anaconda. Heinze's smelter cut the refining costs of the independent mines on the hill almost in half. That was his great material contribution to the prosperity of Butte in the 'nineties.

Independent mines thrived. More independent capital came into the region. Local business and payrolls increased. Heinze was gaining popularity in the district and gaining a reputation extending far beyond the district. This reputation reached into Canada, and Heinze was called to the newly established town of Trail, in the Kootenay region of British Columbia, to build a smelter there for treatment of the Rossland ores.

Leaving C. H. Batterman in general charge of his Butte interests, Heinze journeyed to the new frontier. Capital was available for the building of a small smelter, and Heinze built it.

Owners of the Le Roi Mine, including numerous Canadian Pacific Railway stockholders, signed a contract to supply the new smelter with six hundred tons of ore a day, with a penalty for failure to maintain that production. Heinze built a narrow-gauge railroad from his smelter at Trail, to Rossland. He had learned from observation of Clark and Daly that a newspaper was an invaluable aid in promotional work. Rossland boasted a newspaper, the *Miner*, the only one in the district. Heinze bought it and established P. A. O'Farrell as editor. With this organ he proceeded to arouse public opinion against the Canadian Pacific, bitterly criticizing the inadequacy of its railroad development and service. With the public generally for him because of his practical work in building up the profits of the mines and the general prosperity of the district, it was a simple matter to obtain a land grant from the Dominion and a right of way for an extension of his own short narrow-gauge line to Victoria.

Not until then did the Canadian Pacific awake abruptly to the menace of a driving young American in the conservative British field. Having awakened, it was prompt to act. It asked Heinze his price on all his holdings, and did not quibble when he put that price at one million two hundred thousand dollars. He had something of real value to sell.

The deal was made. Now Heinze had capital in his own right, as well as an established reputation for business shrewdness, promotional achievement, and mining ability which some supporters declared to be little short of genius. He was prepared to make war upon the vested interests of "the richest hill on earth." He did so. And how!

CHAPTER XI

HEINZE, CLARK, DALY

AGAIN in Butte, Heinze discovered that the news of his achievements in Canada had preceded him. Where he had been looked upon as a rising young man he was now accepted by almost everyone except Daly and Clark as having arrived. He was now a capitalist as well as a worker.

But there was a fly in the ointment. C. H. Batterman, whom he had left in charge of his Butte interests, had resigned and gone over to the Boston and Montana Company as consulting engineer. Incidentally he had taken many of Heinze's maps and notes bearing upon the geological formation of the Boston properties with him. Not so good!

These data had been collected in the first months of the young man's surveying experience in the Boston and Montana's workings, and supplemented by his own exploration underground in the Rarus, the Estella and the Glengarry. They had convinced the youthful engineer that the apex of the Boston vein was not in their own ground. Probably it was to confirm that theory that he had taken an interest in the Rarus, mined there for a year and, despite lack of profit, had paid two hundred thousand dollars for a half interest. The fact that later he paid another two hundred thousand dollars for most of the remaining shares seems to substantiate that belief. With complete control of the Rarus and cash enough to work it, Heinze quickly established the accuracy of his apex theory to his own satisfaction.

A brief digression must be permitted here to explain that "apex theory" to readers interested in the story but unfamiliar with mining terms and laws. For upon that theory Heinze based a war that kept Butte and Wall Street in a turmoil for ten years, gave millions to counsel, and eventually forced even Standard Oil magnates to bow to his terms in millions.

The federal mining laws, dating back to 1866, were designed to protect the prospector who first located an outcropping mineral vein. Such surface indication of valuable ore was known as the apex of the vein. The owner was guaranteed the right to follow that vein downward, even when it led under the surface holdings of claims located beside it. That would have been fine if veins were always continuous from the surface down, but too frequently they are not. They are broken or faulted, cut off here and elsewhere by worthless rock. If a vein leading down from the surface is lost near the vertical side wall of a claim, and a similar vein of identical ore is found below it or to one side in the adjoining claim, who is to decide whether the second discovery is a geological continuation of the first? Who but the courts, basing decision on the expert testimony of geologists and engineers?

Newspaper readers who have scoffed at the array of expert witnesses on both sides of a criminal case in which a murderer's defense seeks to cheat the gallows through a plea of insanity, will have some idea of how easily mine owners representing millions of dollars, with more millions involved in the suit, might find expert witnesses to their claims.

Tens of millions of dollars were spent in litigation on this issue in the effort to determine the apex and consequent ownership of ore bodies in the Comstock Lode. Millions were saved by a consolidation of Goldfield properties to avoid such battles in the courts. No less a man than Bernard Baruch was instrumental in financing that consolidation. Perhaps no better proof of business acumen in the perilous mining game has ever been recorded than the late Senator George Hearst's purchase of twenty-six hundred and sixteen acres of apparently worthless side claims in Deadwood Gulch to preserve the great Homestake Mine from that costly litigation.

No such precaution had been taken in Butte. It was Heinze's opportunity, and Heinze was the boy to grasp it. He sued Batterman for return of his maps, charts and papers, and won the action. Sinking his shaft and extending his exploration work farther and farther into the Rarus he encountered rich ore near the vertical side wall of the claim and followed it through that side wall into the adjoining Boston and Montana property.

There the ore was even more valuable, and Heinze proceeded to stope it out, hoist it and rush it to his smelter as fast as possible, working twenty-four hours a day. The Boston and Montana smelled a rat, sent spies into the Rarus, and decided that Heinze was stealing their ore. When they called for an explanation Heinze asserted that he was merely following the vein which apexed in his own Rarus and ran down through the side wall of that property. Perfectly legal, if true. But was it true? The Boston and Montana said not, and brought suit to enjoin Heinze from further mining through the Rarus shaft.

Heinze was shocked and grieved, but offered to avoid controversy by paying the Boston and Montana two hundred and fifty thousand dollars for the disputed property which, he said, had cost the company only twenty-five thousand dollars, and which had been allowed to lie idle for years until his energy and expenditures had found the ore. But the fact remained that the ore had been found, in such quantities and of such high grade that competent engineers estimated its value at twenty million dollars. The brazen effrontery of Heinze's offer of two hundred and fifty thousand dollars for property apparently worth twenty million dollars added insult to injury.

The Boston capitalists, since the days of their development of the Calumet Company in northern Michigan and the consequent flooding of the Back Bay with endless millions in dividends, had rather fancied themselves as kings of the copper world. They had been a little slow in getting a foothold in Butte. William A. Clark and Marcus Daly had beaten them to the good properties while those properties were cheap. They had met with some difficulties in trying to operate in Butte in the grand manner, as they had operated in the Calumet and Hecla Mines. They were a little peeved. But that irritation was as a mosquito bite to a lion bite compared with the effect of Heinze's casual offer of two hundred and fifty thousand dollars for twenty million dollars' worth of ore.

A. S. Bigelow, dignified president of the Boston and Montana, was almost beside himself with rage. The atrocious impudence of this upstart engineer, who only a few years earlier had been working for him at five dollars a day in that same Boston and Montana Mine, left him gasping. But not for long.

According to a report of the incident published some years later in a magazine of national circulation, Heinze had taken his maps and charts to Boston and spread them out on Mr. Bigelow's broad mahogany desk in support of his contention that the Boston ore apexed in Heinze's Rarus Mine. Bigelow swept them aside and informed the upstart that Boston had already been badly treated in Butte, and would stand no more such treatment. The Heinze threat was blackmail. The Heinze proposal was an insult. Boston must discipline someone in Montana, and the decision was taken therewith that it would be Heinze.

"Very well, Mr. Bigelow," Heinze replied politely, "you have a great deal of property in Montana which is subject to the same kind of litigation as that which you say you will thrust on me. If your program is fight, you will find I am prepared. Before you and I have finished, I will give you a fight that will be heard of from one end of this continent to the other."

Heinze was accurate in that forecast. Three years later he announced to the world that Bigelow's private fortune of ten million dollars had been destroyed and his Globe National Bank wrecked. But the Boston and Montana and the Butte and Boston companies of which Bigelow was president had not been destroyed. They inherited the war, and the war raged on.

Probably shocking, but not demoralizing, was Heinze's discovery that Bigelow had been eliminated as commanding general of the opposing forces only to be succeeded by the vastly more powerful generalship of Henry H. Rogers and William Rockefeller of the Standard Oil group, with James Stillman of the City National Bank in the important post of quartermaster-general. But Hercules, striking off the heads of Hydra, only to see two grow where each one had grown before, found a way out of that embarrassing situation.

The battle with Bigelow was merely a training bout for Heinze. He went back to Butte from his first interview with Bigelow, and took inventory of his resources. He was in good standing with the miners, whom he had always paid well and treated fairly. He stood high with the independent operators, whose smelting costs he had cut almost in half. He was popular with the local business men whose trade he had increased with his stimulation of independent mining. With that support he obtained the nomination of William Clancy to the bench of the District Court. Clancy was elected. Heinze was set for any action the Boston and Montana or other companies might bring against him, or for any action he might decide to bring against them.

Clancy was a curbstone lawyer of benevolent appearance, but little education. His outstanding trait was stubborn loyalty to his friends, and complete indifference to his opponents. He was coarse and vulgar, but cunning and shrewd on occasion. Although so many of his judicial decisions favored Heinze that he was popularly supposed to be "owned" by Heinze, John J. McHatton, Heinze's counsel in many of these cases, still insists that that was absurd.

In evidence, Judge McHatton cites the fact that several of the most important of these decisions were upheld by the State Supreme Court on appeal. Also, according to McHatton, Clancy occasionally took action in direct opposition to Heinze's interests.

On the bench Clancy was a burlesque of judicial dignity but he was no joke. With feet cocked upon a chair or upon the desk before him, he would gaze out the windows of the courtroom while the most highly paid attorneys in the country and the most famous geologists and engineers of the world wrangled to wedge their evidence into the record. From time to time he would turn his head just far enough to add a generous portion to the pool of tobacco juice in a huge brass cuspidor conveniently placed. The arguments of counsel or the testimony of witnesses left him unmoved. Indeed he seldom appeared to be hearing what was said.

Only when an objection was made would he appear to come out of his ruminating trance long enough to sustain it if made by Heinze's counsel or overrule it if made by the opposition. When the wrangling reached a point which disturbed his calm he would announce brusquely: "I've heard enough of this. Call the next witness." It made no difference to Clancy whether the witness so dismissed were a man of the caliber of John Hays Hammond or a mucker in the mines.

Off the bench, Clancy's lack of dignity and slipshod personal habits made him popular with miners, bartenders and the lower level of Butte society in general. Their votes, shrewdly advised by Heinze, kept him on the bench. They delighted in stories of his crudities. Here was no high and mighty dignitary, but a man of the people, one who could be slapped upon the back in a barroom, one who might buy a drink for an admirer, one who would safeguard the interests of the common man from the capitalistic grasp of the plutocrats.

Stories of Clancy's unassuming and unpretentious manner grew and multiplied. The coarser the incident narrated, the more popular. An acquaintance meeting him on the street one day stopped for a moment's chat. "Judge," he said, "I'll bet you a dollar I can tell you what you had for breakfast this morning."

"Take you," said the judge genially.

"Ham and eggs." The acquaintance lifted the judge's long white beard and displayed the stains of egg yolk and ham fat. "You lose," said the judge. "That was yesterday's breakfast."

Butte delighted in that story. Butte was that way. Clancy was that way. Everything was favorable to Heinze's promise to give the Boston and Montana Company a fight that would be heard of from one end of this continent to the other.

In further preparation for that struggle Heinze again studied his maps, rechecked surveys of the surface lines of the claims on Butte hill, and discovered a triangular fraction of the hillside containing about forty square yards which had never been recorded or patented. The land lay in the area between the Anaconda, St. Lawrence and Neversweat shafts. Heinze instantly seized it, secured title, and named it the Copper Trust, announcing that the inexhaustible bodies of rich ore in those three great mines all apexed in his tiny fraction.

With that he had incurred the enmity of Marcus Daly, the largest operator in Butte with the richest and most extensive mines. Daly promptly lined up with the Boston and Pittsburgh capitalists against the Heinze menace.

Only William A. Clark among the leading powers of Butte remained, apparently, friendly to the young engineer. At the moment Heinze claimed no apex of veins running into the Clark mines. Clark was not a fighter of the Heinze or Daly type. Always he preferred compromise to violence. Consistent devotion to that policy, coupled with an unremitting energy and some luck, had proved highly profitable. He was perfectly willing to see Heinze fight the Boston and Montana, and quite delighted to see him fight the Daly interests. Those battles would cost Clark nothing, and might be of value in revealing authentic data concerning the geological formation underlying Butte hill and the Clark mines.

Clark had found the waiting game paid well in Butte. His first demonstration of that had been in the Moulton Mine, high on the hilltop west of the Alice. He had allowed that property to lie undeveloped for five years until the Alice shafts and drifts, without cost to him, had indicated that the Moulton contained large bodies of rich ore. Then Clark had gone to work on that property, opened the ore bodies where the Alice exploration had suggested they might lie, and taken out enough silver ore to justify erection of a forty-stamp mill just north of Walkerville in 1881.

Continuing his development of the mine, he had timed his sinking of shafts and had placed his crosscuts always on a level above the adjacent workings of the Alice. This permitted the water of his mine to seep away into the deeper diggings of his neighbor, where the other company would be forced to pump it out without cost to Clark. Independent mining engineers of Butte who at times did work for Clark as well as Heinze, Daly and other operators are authority for that statement. Arthur V. Corry is one who asserts that this shrewd practice, combined with a considerable amount of luck, was an important factor in Clark's success. In substantiation of the element of luck he cites Clark's profits from the Mayflower Mine.

"Clark never had the confidence of Marcus Daly in the inexhaustible riches of Butte," says Mr. Corry. "He never had the fighting spirit of Heinze. He was smart, cautious and watchful. Also he was lucky. It was proved beyond question in the Mayflower. When the workings were down a few hundred feet they ran out of ore. Clark ordered the mine to be shut down. His superintendent begged for another thirty days to find a continuation of the ore body. Clark insisted that it was useless. The ore had been exhausted, not merely lost. But finally he gave in, and agreed to pay the bills for another thirty days. Even so the mine would owe him nothing.

"The superintendent pushed his exploration work to the limit, and in less than the allotted time opened up another stope that produced two million dollars. When that was exhausted Clark again ordered the mine closed, and the machinery removed. Again the superintendent protested, but this time in vain. Clark didn't believe in luck. The fact that it seemed to have come to him once made it all the more unlikely that it would come again. He allowed the superintendent a limited time to take out the machinery and clean up the mine for closing.

"Perhaps the superintendent did not believe in luck either. But he believed in the mine, and in his own nose for ore. Instead of cleaning up the mine he put his force to the following of a new lead. And before the day set for closing came around he had entered another ore body from which two million five hundred thousand dollars was taken to be added to Clark's fortune."

So Clark was willing to see Heinze finance the great legal battles against the Boston and Montana and against Daly's St. Lawrence and other properties of the Anaconda group. By this time Clark had extended his copper and banking interests to the vast newly discovered copper deposits of Arizona, and was well on his way to the sixty million dollar profits which he was to take from the United Verde Mine at Jerome.

His political ambitions had soared anew with his triumph over Marcus Daly in the battle to establish the state capital at Helena. F. Augustus Heinze, who was developing a talent for public speaking of a character calculated to win votes, looked like an able lieutenant. Heinze's local success in putting the disreputable William Clancy on the bench was further evidence that he might be valuable to the would-be senator. Clark patted him on the back and told him to go to it.

In the meantime Clark would appear as a benefactor to Butte, where there was still some resentment against him as a result of the capital fight. He planned an outdoor recreation center for the smoke-ridden city. Columbia Gardens, in a canyon of the range two or three miles east of Butte was laid out with playgrounds for the children, a dancing pavilion for the young people, picnic tables and various concessions for the families. Clark's street railway was extended to the spot to provide transportation, and incidentally to bring some cash return to Clark.

The miners, mill workers and other residents of Butte who could not afford recreational trips to New York, or even the summer camping and fishing expeditions of the well-to-do citizens in the sportsmen's paradise of the Rockies, flocked to

Columbia Gardens. Clark financed band concerts, improved the recreational facilities, and basked pleasantly in the approval of his fellow citizens and voters. His personal stock soared in Butte.

Years later, when the devastating blight of sulphur and arsenic smoke from the roasting ore heaps and smelters of Butte had been eliminated, when automobiles had opened the surrounding mountain playground to the workers, when flowers and trees had added to the attractions of the playground, we find Clark quoted: "The Columbia Gardens is my monument. Of my many business enterprises it is the one I love best, and it is practically the only one on which I lose money."

Probably he did lose money, but by that time he had plenty to lose. Unquestionably he gained popular favor, of which he had not enjoyed a great proportion in his Butte career. He was still a little too formal, too cold and calculating, too neat and polished to be a popular hero in a mining camp. Butte's populace was not greatly thrilled by dispatches from New York or Paris announcing that William A. Clark had succeeded in buying Rembrandt's *Portrait of a Man,* Corot's *Dance under the Trees,* or Raeburn's *Portrait of the Artist's Daughter.* Butte preferred the news that "Marcus Daly walked into the Hoffman House in New York last night with $27,000 cash in his pockets, his winnings on his colt Tammany which defeated Lamplighter at Gutenberg."

Art was all right. Butte even approved of it when it appeared in the form of nudes above the back bars of its favorite saloons. But horse racing was far more entertaining. Clark might give them culture, but Marcus Daly gave them thrills. Clark might be spreading the name and fame of Butte through the drawing-rooms and galleries of New York, London and Paris, but Marcus Daly was spending his money in Montana, where he was making it. His ranch and stables in the Bitter Root Valley were something to see.

There was in Daly, especially with reference to his ranch, stables and race-track activities, something of the colorful character of "Lucky" Baldwin. No stables were too good for his horses. No horses were too good for his stables. When his trainers pointed a horse for victory, that horse gave everything it had to the race. But when Marcus Daly smilingly let the tip-seekers into the inside information that this horse or that was out to win, and would win in a walk, the educated money went down on another of the bangtails.

"Daly was a fine fellow," says Ben Stack, who knew him intimately from the day of his arrival in Butte, "but you couldn't believe a word he said about a horse race."

"When Mr. Daly said one of his horses would win, wise people were careful not to bet on it," says Mrs. T. J. Murray.

But Butte followed the Daly horses with enthusiasm and profit, outside of his tips, because they knew the quality of his stables and trainers. Horses which are still named in fond reminiscence when grizzled old drivers get together came from his stables. There were brood mares with reputations which will never be forgotten among the devotees of the trotting race as the highest form of competitive sport. There was Rachael, 2:08¼; Ellorce, 2:08¼; Red Silk, 2:10. Famous stallions included Prodigal, 2:16 to an old-fashioned high wheel sulky; Ponce de Leon, 2:13.

Of the running horses perhaps the most famous was Tammany. In his first race on the eastern circuit Tammany romped home at sixty to one. Butte had not only cause to celebrate but the cash with which to do it. Tammany won all his stake races as a three-year-old, and defeated the famous Lamplighter as a four-year-old. On his way home from that victory he contracted lung fever, and a large part of Marcus Daly's twenty-seven thousand dollars won on the race went for veterinarian skill and the champagne and porter with which the horse was dosed until he recovered. Butte liked that story also.

The town celebrated in characteristic style when Daly's Montana won the Suburban at forty to one. Then it roared with laughter when it discovered that Marcus Daly had had a one thousand dollar ticket on his horse at those odds, had left the ticket in his clothes in "the dry" when he had been called into one of his mines on the day of the race, and had returned to the surface to find "the dry" burned up with his clothes and the ticket worth forty thousand dollars in it.

Even the personality of Daly's horses was interesting. Tammany was a pet, playing with the fox-terrior puppies which shared his stall, gently nuzzling children who strayed into his paddock. Montana was an ugly awkward brute, winning his races by main strength and meanness. He terrorized every groom who was forced to come within reach. On one occasion he broke two tie reins while being groomed, seized the unfortunate hostler by the shoulder and carried him around the stable yard until rescued, with a crushed shoulder. One of the brute's favorite amusements was the killing of rats which might appear in his stall or attempt to rob his feed box.

Marcus Daly understood horses and men as well as mines. He bought the best, and hired the best talent to improve his property. When his great stables in the Bitter Root boasted such animals as Lady Reel, Hamburg, Montana and Tammany, Daly engaged Jim Lucas, the famous trainer who had conditioned some of August Belmont's best horses for their greatest races. Lucas had been famous even before that, having won a reputation on R. A. Alexander's Woodburn Farm in Kentucky. Under the direction of Lucas in Daly's employ the great Ogden was bred and trained to fame and profit, to become the grandsire of Zev. Zev won the Kentucky Derby in 1923, and with Earl Sande up defeated England's hope and pride, Papyrus, in the one hundred thousand dollar International.

When Daly wanted an addition to his stable he bought it, regardless of the price demanded, and frequently made money on the deal. He paid forty-one thousand dollars for Hamburg, won sixty thousand dollars with him, and sold him for sixty thousand dollars. He paid fifteen thousand dollars for Lady Reel, in foal with Hanover; sold Hanover for ten thousand five hundred dollars, and sold two fillies as weanlings from the same dam for fifty-four thousand dollars. When his stable was sold after his death it brought seven hundred and fifty thousand dollars to the estate. August Belmont's sale of one hundred and forty head had grossed only five hundred and fifty thousand dollars.

Butte liked all that. But most of all it liked the summer meetings of the Montana Jockey Club of which Marcus Daly was president. All the youth and beauty, the age and ugliness, the virtue and vice of the town crowded grandstands, boxes and rail for the spectacle. Sons and daughters of well-to-do citizens, smartly dressed, home from college or from a gay winter season in New York, filled the clubhouse veranda or the boxes with life and color. Daly gave his miners a day off, with pay, to attend.

Madams from the town's most rococo parlor houses displayed themselves, their handsomest girls and their Paris raiment, and gave back to the bookies some of the cash left with them by the sporting element so attracted. Miners, mill men and others off shift in a community working twenty-four hours a day crowded the betting ring and the bars between races, and lined the rail in front of the grandstand for the colorful parade and the flashing speed of the great horses. Race days were great days in Butte.

The ugliness and grime of the city were forgotten. Even vice lost its sordid aspect when clothed in silks and jewels The bands played, the horses pranced and flashed like light, the sweating toil of thousands of men within the earth was forgotten. Even personal enmities and jealousies seemed to be laid aside.

Heinze, Daly, the Clark boys, Will and Charlie, mingled in apparent amity on the clubhouse veranda. Only William Andrews Clark of the leading citizens of Butte seldom found time to be present. He was still engrossed in the business of piling up more millions, and perfecting his organization for another bid for a seat in the United States Senate.

CHAPTER XII

POLITICS AND BRIBERY

BUTTE, and indeed all of Montana, was a strange mixture of culture and depravity by the time William A. Clark began his second campaign for election to the United States Senate in 1898. Butte's theaters were filled to the last seat in the topmost gallery for the performances of such artists as Ward and James in Shakespearean repertoire. When the theaters were closed by a war of managements Sarah Bernhardt played in an old skating rink on Montana Street, and packed it to the doors.

At the same time the girls "rustled the boxes" of the Comique. Men who might have been expected to be at a concert of Emma Nevada or a performance of Melba were stuffing stockings with the red percentage tickets accompanying their purchases of champagne behind the screens. The younger generation with no reputations to maintain was pinching the legs of complacent "entertainers" circulating among the tables. The population of the segregated district was approaching the three thousand which it ultimately reached.

"Swede Sam" Mullin was banking a game in the Butte Hotel in which $29,295 changed hands in nine turns of the wheel.

The Silver Bow Club had arisen to the dignity of entertaining Henry H. Rogers of the Standard Oil Company, and J. P. Morgan, of the house of Morgan. The petty gambling game of Panguingue, according to the description of some contemptuous old gamblers, had been adopted "to keep the pimps off the street."

Dublin Gulch had become so tough that when the elder J. P. Morgan was taken through that famous thoroughfare in the open hack of "Fat Jack" Jones, the Morgan dignity was plastered with an overripe tomato. A mild bit of playfulness on the part of the Dublin Gulchers, and one which Mr. Morgan was about to resent with violence until "Fat Jack" persuaded him that if he set foot in the street with any such intention he probably would be beaten within an inch of his life. Dublin Gulch was no respecter of persons. "Fat Jack" knew it. He himself, tall and lean, with a long goatee and high silk hat, was Butte's most familiar figure to all travelers. He was jehu to all the notable visitors through many years. Between times he might be called to drive for Lou Harpell, Mabel Loy or Belle Rhodes on a shopping tour. Or again, and quite frequently, he might be found, silk hat and all, dropping his fare on a faro layout while his cab horse sadly awaited its oats in the street.

Clancy could be a judge. Max Fried could be a ward boss. Charlie Clark could stop the show when he entered a box in the theater after an absence in New York. Gertrude Atherton could be feted, and later hated, when she wrote a novel based upon the local color of Butte. Mary MacLane could stir the literary critics of New York and Boston with her "peripatetic philosophy" and her youthful dissertations on sex impulses, and shock the spinsters of Butte with a casual question: "Have you ever been seduced?" Mary MacLane could even be pictured by one portrayer of Butte's versatility as standing on Butte's Broadway with a basket of groceries on her arm calmly watching "Callahan the Bum" trying to commit suicide by hanging himself to the awning rope of a jewelry store.

A single Butte drugstore could sell hundreds of dollars' worth of drugs to addicts each month. Leading business men and politicians could maintain mistresses without losing social or political standing.

That was Butte when William A. Clark began his campaign to elect a state legislature in 1898 which in turn might elect him United States senator in 1899. Even the United States' declaration of war against Spain could not make much of a sensation in that field. Some mining interests took the war seriously in the belief that it would mean an increase in the demand for copper and a consequent increase in the prosperity of Butte, but Montana was more interested in intrastate than in national or international affairs.

According to William A. Clark himself, "the conditions were such in Montana—everything in the way of elections there, where this company (the Anaconda) has had domination, has been of such high-handed character—that nobody could expect to have any recognition whatever unless he bowed the knee and crawled in the dust to these people. The primaries at these elections held in the city of Butte, where the delegates to the convention were to be selected, showed that in one or two precincts at least the total of votes cast was about double the number of voters in the precincts. And those were the precincts controlled by the Anaconda company. I was not present when that election occurred, but I know positively through my son and hundreds of others who have told me about it . . . that there were repeaters traveling around from poll to poll in the city of Butte, and it is said that a train was run from Anaconda to South Butte with a load of voters and that they were taken around there and voted at several polls. They own the political machinery of the county, or a good deal of it. I saw men at the election receiving money for votes at that election. . . ."

Although that was sworn testimony by Clark before an investigating committee comprised of seven United States senators, aided by six distinguished lawyers, there is no evidence in the record that either the senators or the lawyers noted its outstanding contradiction: "I was not present when that election occurred," and "I saw men at the election receiving money for votes at that election."

Perhaps that is a minor point. Certainly it may be taken for granted that politics in Butte and throughout Montana were thoroughly corrupt. Sworn testimony is available that one million, two-hundred and fifty thousand dollars was expended in the political battle to make Helena rather than Anaconda the state capital in 1894. That would have been twenty-five dollars for each of the fifty thousand voters then registered in the state. It will not be contradicted either by historians or by residents of Montana who were active in the 'nineties and who have contributed to the information in this record, that political control was clearly in the hands of Marcus Daly when the state gave a plurality of more than thirty thousand out of a registration of little more than fifty thousand to William Jennings Bryan in 1896.

Neither will it be denied that the influence of Daly, the Irish mine boss who had built the largest fortune in Montana with the exception of Clark, was most important in Butte and Anaconda where his companies employed a majority of all wage-earners.

Daly had not been above borrowing a chew of tobacco and discussing politics with a laborer when he was an almost unknown quantity in Butte. He spoke their language, with all its "dones" and "ain'ts." He was not above it when he employed five thousand men and had accumulated a fortune running into many millions. Testifying at the same hearing in which Clark made the statement quoted above, Daly replied to a question as to whether he had ever used any means to coerce voters by saying, "I never had to."

In these circumstances there was no doubt that the Democrats would be overwhelmingly victorious in the state election of 1898.

There was a doubt in the mind of Clark, however, that it could be made overwhelmingly Daly-Democratic. He yearned to free it from the control which he characterized as corrupt and demoralizing. He yearned so strongly that he put up one hundred and thirty-nine thousand dollars for the secret uses of his private campaign committee against a total of seven thousand, seven hundred and fifty dollars contributed by Marcus Daly to the official Democratic organizations.

Evidently Clark's assertion that Daly ruled was accurate. Even with expenditure of one hundred and thirty-nine thousand dollars without accounting, Clark's private committee was unable to elect a single candidate to the legislature from his own county of Silver Bow and his own city of Butte. The Daly Democrats went in with a whoop. Less than one-fourth of the legislature was Republican. Less than one-fourth of the Democrats were pledged to the support of Clark.

Clark's campaign to "purify" the Montana Democracy by the expenditure of one hundred and thirty-nine thousand dollars against Daly's expenditure of seven thousand seven hundred and fifty dollars to keep it pure Daly had been a failure. To some it was a joke. Witness an extract from a subsequent examination of Daly under oath:

"Mr. Faulkner.—'Where you and Clark live is Democratic, is it?'

"Mr. Daly.—'We live at the same place. That accounts for it going Democratic. It keeps us all our time purifying them."' Prior to that, Daly had had a conversation with Frank Corbett, a prominent and influential resident of Montana, concerning the situation in the newly elected legislature. The conversation, as narrated, took place in New York in December, 1898. Said Daly:

"He [Corbett] met me one afternoon and asked me if I was going back to Montana. I told him I was not. He said: 'Ain't you going to take part in this senatorial fight?' I says: 'No, sir; I have got through with that pool.' He said: 'Mr. Clark has just gone out there, and,' he says, 'I understand he is going to be a candidate, and they think he will make a pretty hard fight.' I told Mr. Corbett—I said: 'I don't think Mr. Clark will be a candidate. I don't think he will have anything to do with it. He was hopelessly defeated at the polls. He hasn't got a single vote from his own county, and it would mean wholesale bribery, and I don't think Mr. Clark would want to go into that.' That was all. The conversation dropped then. I asked him in return: 'Are you going out, Mr. Corbett?' He says: 'No, sir; I don't want to have anything to do with it.'"

That was the situation in which the Montana legislature met on the second of January, 1899. Election of a United States senator was its most important task. The only other business before it was the passage of two bills which, strangely enough, had the support of both the Daly and Clark factions. The pleasure of a political fight was never allowed to stand in the way of potential financial profits. In the greater excitement of the senatorial contest the bills were pushed through without difficulty. One of them was to be of vital significance in the subsequent history of Butte. It will be considered in its place.

In the meantime money, gossip and scandal were floating around the legislative halls, the hotels, the clubs, streets and saloons of Helena. The easy money of the capital campaign five years earlier had not been forgotten. The *Anaconda Standard's* charge, made near the end of Clark's senatorial campaign a year before the capital fight, had not been forgotten. That printed story had stated: "Clark will be elected if money can buy his election, but in no other way."

There were members of the legislature who had had to borrow traveling expenses to reach Helena. There were members who owed board bills in their home towns. There were men whose financial liabilities exceeded their assets. There were men in danger of losing their small properties through foreclosure. There were men who had been living on seventy-five dollars a month and less. It seemed a fertile field for the sowing of dollars and the reaping of votes.

The committee which had spent one hundred and thirty-nine thousand dollars of Clark's money, without accounting, in the election of that legislature was now established in Helena. Charles W. Clark, son of the candidate, was its treasurer. William McDermott, A. J. Davidson and John B. Wellcome, who had become known as members of that well-financed committee, were all to be seen in the hotels, bars and the Helena Club.

W. G. Conrad, with the support of the Daly Democrats, was the leading candidate for the senatorship on the first ballot, with William A. Clark some distance down the line. Neither candidate nor any other of the half-dozen voted upon in the first few roll calls had anywhere near the necessary majority. More liquor flowed in the bars, more whispered conversations took place in the lobbies, more talk of the cash value of votes was heard in the streets. A deadlock such as had prevented the election of a senator in 1893 when Clark had also been a candidate again threatened. With that threat the curb quotations on votes went up swiftly. Unofficial charges were freely made that Clark was buying votes.

To sell, or not to sell; that was the question. There was no doubt in evidence that the buyers were at hand with the cash. And in that electric atmosphere State Senator Fred Whiteside asked the speaker of the joint assembly for recognition. Rumors had spread that this was to be the big day. The galleries were crowded. Legislators and spectators were on the edges of their seats. Whiteside brought the drama to a climax as sensational as anyone could wish.

Producing four envelopes, he extracted from them thirty thousand dollars in currency—more one-thousand-dollar bills than anyone in the auditorium had ever seen before—and announced in ringing tones that this money had been placed in his hands by agents of William A. Clark to buy the votes of himself, W. A. Clark of Madison County, H. L. Myers and H. H. Garr.

The chamber was in an uproar. Newspaper reporters tumbled over one another in their eagerness to get the story on the wires. Montana was ready to believe any rumors or gossip of bribery, but a definite and official charge, made upon the floor of its legislature and backed by evidence in the form of thirty thousand dollars in currency, that its wealthiest citizen had actually provided that money to buy his way into the United States Senate, was almost more than it could believe. However, there was the thirty thousand dollars.

No justice can be done to either side in the controversy or to the journalism of the day without reprinting the headlines of the two leading newspapers of January 11, 1899. The *Anaconda Standard* rushed to the streets with a great black banner line, and subheads as follows:

CLARK BRIBERS CAUGHT AT IT RED HANDED

THIRTY THOUSAND DOLLARS OF THE BOODLE PILED UP IN FULL VIEW OF MEMBERS OF THE LEGISLATURE—ASTOUNDING REVELATION OF HIGH CRIMES AGAINST STATE

In the detailed story which followed, no less astonishing to modern newspaper readers who are accustomed to some editorial discretion in stories which might be construed as libelous, Clark was referred to as "the Paris millionaire." The *Standard's* editor knew as well as the demagogues of a later day how to arouse the resentment of laboring men against a man of refinement. The *Butte Miner,* Clark's own paper, in entertaining contrast to the *Standard,* headlined the story as follows:

A DAMNABLE CONSPIRACY

DALY CROWD SPRINGS THEIR PROMISED SENSATION—BUNGLING WORK AT THE OUTSET

STATE SENATOR WHITESIDE, NOTORIOUS FOR HIS CONNECTION WITH THE CHARGES AGAINST THE STATE CAPITAL COMMISSION, THE TOOL USED BY THE GANG WHO ARE ATTEMPTING TO RUN THINGS AT HELENA ON THE ANACONDA PLAN—ABORTIVE ATTEMPT TO STAMPEDE MEMBERS OF THE LEGISLATURE BY EXHIBITION OF MONEY AND CHARGES OF BRIBERY—WHOLE THING BEARS EVIDENCE OF HAVING BEEN COOKED UP BY THE ARCH-CONSPIRATORS FOR USE ON THE DAY WHEN THE FIRST BALLOT FOR UNITED STATES SENATOR WAS TO BE TAKEN—RESULT OF THE VOTE SHOWS THAT THE SCHEME DID NOT PAN OUT AS WELL AS EXPECTED.

Elsewhere in the *Miner,* under a two-column headline—"It Was a Daly Trick"—an editorial explained that the charges were made and the thirty thousand dollars in bills provided to support them in an effort to cause such confusion and excitement that party lines would break up and allow the Daly forces to rush their candidate through to victory.

Certainly there was confusion and excitement, but no candidate was rushed through to victory. In the vote of that day Conrad polled thirty-eight, Toole twenty, Marshall fifteen, Clark ten, Hartman four, Frank one. Clark had less than one fourth of the votes necessary to elect; Conrad, candidate of the Daly forces, had more than three-fourths.

Helena, Butte and all Montana seethed with excitement. Newspapers throughout the country gave the story recognition as front-page news of national interest. For a time it ranked above the "embalmed beef" scandals of the Spanish-American War as a subject of importance. It was accorded far more space and comment than the hanging of twenty-one highwaymen by the Vigilantes in thirty days had been given only a generation earlier. "Civilization" in Montana was advancing.

The *St. Paul Dispatch* published a three-column cartoon showing William Jennings Bryan collapsing before a screen on which was depicted the situation in Montana. The picture displayed a package of one-thousand-dollar bills with the caption, "Credentials of the coming Democratic Senator of Montana;" a thousand-dollar bill couchant, with six-shooter and bowie knife rampant, as the new coat of arms of Montana; a thousand-dollar bill with the caption, "The kind of bill most frequently introduced in the Montana legislature;" a row of thousand-dollar bills as a "Frieze on the walls of the Montana legislature;" one legislator lighting a cigar with a thousand-dollar bill; a group of legislators entering a door labeled "Bribery Room." Gasping with chagrin, Bryan collapsed at the spectacle with the remark, "Toö much prosperity there for me."

Montana was in disgrace, but either failed to realize it, or was indifferent to it. That William A. Clark was not in disgrace was indicated in the next day's ballot when his vote leaped abruptly to a new high of twenty-one. Of this addition, however, not one came from Conrad's support. All the new votes for Clark on that day came from Marshall's column, which dropped to zero. It was evident that considerable numbers of the legislators were as ready to believe Clark's charges of conspiracy as they were to believe Daly's charges of bribery.

Senator Whiteside, however, had his story and was sticking to it. The clerk of the assembly had the thirty thousand dollars in bills with the four marked envelopes in which they had been contained. These were turned over to the state treasurer to hold as material evidence. As the days went on and Clark's support grew from twenty-one on January twelfth to twenty-five on the sixteenth it became evident that the "exposure," as it was called, was not shocking either the legislature or the electorate of Montana half as much as it was shocking the rest of the United States. Montana had already been demoralized by easy money, open gambling, quick and unearned fortunes, low moral standards in the proletariat, subsidized newspapers, complacent representatives of the law. It might howl about bribery and conspiracy but it was not shocked.

Charles W. Clark, John B. Wellcome, William McDermott and A. J. Davidson were still in Helena and were still recognized as the agents of William A. Clark. Also they were known to have far more resources than the thirty thousand dollars in bills alleged to have come through Wellcome in payment for four votes. They remained in even better standing than State Senator Fred Whiteside who had accused them and who, in turn, had been accused by the Clark forces as being the agent of the Daly interests in a plot to destroy Clark by scandal. No one with millions could be destroyed by scandal in the status of civilization which Montana had attained in 1899.

Still the legislature appeared to be deadlocked. Gossip in the lobbies and on the street corners reported that the price of votes was going up. It did not reveal any popular reaction against the alleged bribers, or the alleged conspirators. But something had to be done about it, not because the electorate was shocked but

Montana Historical Society, Helena

Bills in marked envelopes—Daly's conspiracy or Clark's bribery?

COURTESY OF MONTANA HISTORICAL SOCIETY, HELENA

Copper processing: Under smoke-gray skies, black-smudged miners turn reddish-brown ore into crisp green bills. (Photo of Central Butte Hill and Dublin Gulch circa 1905.)

because the days were passing. The session of the legislature was limited by law to sixty days, and there was danger that the state would be left without a junior senator as it had been by the deadlock of 1893.

Whiteside, backed by a group of legislators, went before the grand jury with his witnesses and evidence. Attorney-General Nolan presented the information. That appeared politically dangerous to the Clark interests. They countered by bringing charges before the legislature that Whiteside himself had been elected to the State Senate by fraudulent counting out of some of the votes for his opponent.

Whiteside, according to a synopsis of evidence before the grand jury later made public by Attorney-General Nolan, testified as follows:

"When Clark (of Madison County) agreed to help I talked the matter over with him and told him that I wanted him to agree to take the money which they had authorized me to offer him, and to insist that it should be put in my hands. I saw Mr. Wellcome and told him that Clark would accept the money, and he said he would like to see Mr. Clark and have a talk with him. Mr. Clark and myself went to Mr. Wellcome's room. . . .

"Mr. Wellcome went over to one corner of the room and made some remark about a picture, and beckoned to Mr. Clark, who went over there, and the money, $10,000, was put into an envelope and the envelope sealed and delivered to me. Mr. Clark and Mr. Wellcome seemed to count the money, and the money was in an envelope, and after being counted Mr. Clark wrote on the back of the envelope. I said something about wanting to understand what the conditions were, and when the money should be turned over to Mr. Clark. Mr. Wellcome said that Mr. Clark, of Virginia City, was to vote for W. A. Clark for Senator, 'so long as we want, or so long as there is a show of his election. It is not contingent upon his election to the United States Senate. If he votes for Mr. Clark he gets the money, whether he is elected or not.' . . .

"Sometime later in the day I met Mr. Wellcome, and he asked me if Myers was around, and I told him I thought he was in the lobby. He said he would like to fix the matter up. There was some other gentleman in the room, and I said, 'Where can we meet him?'

"'We can get room 206,' he said. 'Take him in there and I will come in.' I went down and got Myers and took him into 206. Went in for Wellcome and he stepped into the bathroom and gave me the money for Myers, and he said, 'You can fix it with Myers. I will trust you with it.' I told Mr. Wellcome I would like to get some money for myself, and he said, 'Do you want the whole $10,000?' I said I would only need $5,000. I didn't care to arouse suspicion by wanting the $10,000, and he gave me the $5,000. . . .

"Mr. Wellcome said they had hustled for large size bills, but the old man—referring to W. A. Clark—was coming from Butte that evening, and that he would bring a supply. When he did come that night, in one of these cases, the money was paid over. . . ."

The testimony of W. A. Clark of Virginia City, Madison County, before the jury, was substantially corroborative of Whiteside's statement. He said: ". . . About half past nine or ten o'clock that night, on January 4th, Whiteside came to me and said that Wellcome wanted to see me in room 201. Mr. Whiteside and myself went to room 201, and we three took seats in a line. After a time Mr. Wellcome got up and went over in the corner and took out a bunch of bills and beckoned to me to come over. Something was said by him in regard to talking there; that Mr. Campbell occupied the room next to us and it was necessary to speak low.

"I took the bills and counted them over, and there were ten $1,000 bills. I placed them back in the envelope, sealed the envelope, and wrote my initials on the flap. I then wrote across the end of the envelope, 'The property of W. A. Clark.' and then took my stylographic pen and threw a large blot of ink on the envelope as an additional identification. This is the envelope. I then took the envelope which I received from Mr. Wellcome, with the money which it contained, and turned it over to Mr. Whiteside. My purpose in receiving this money in this way was to obtain evidence against the crime of bribery, and the money was simply received by me from Mr. Wellcome not with the intention of influencing my vote but simply for the purpose of getting evidence that could be used to prevent bribery. . . ."

Myers and Garr testified similarly, each identifying the envelopes in which their alleged bribe money had been placed with their initials to identify it before it was handed to Whiteside as evidence for the dramatic "exposure" on the floor of the joint assembly. Then the grand jury went into a huddle. Upon its decision might depend an effort to send Montana's wealthiest citizen to the penitentiary instead of sending him to the United States Senate. Every resident of Montana who could read and write was on edge. Every politician and ward heeler in the state knew that Whiteside, Clark of Virginia City, Myers and Garr had testified. Inasmuch as they knew what these same men had told the legislature in open session, they were pretty well acquainted with what had been told to the jurors.

The Daly faction was jubilant. But the Clark faction had not been idle. Their investigation of the credentials of Lee Whiteside as state senator had probed back to the ballot boxes and revealed that some of the ballots for Geiger, Whiteside's opponent at the polls in the election of 1898, had been thrown out on the charge that they had been improperly marked. Charges that Whiteside had obtained his credentials by fraud were then brought before the State Senate. The evidence was displayed. The Senate unseated Whiteside and certified Geiger to his place. William A. Clark's *Butte Miner* exulted in a flaring headline:

A MORAL LEPER LOST HIS SEAT

SENATE CHAMBER WILL NO LONGER BE POLLUTED BY FRED WHITESIDE—VOTE WAS TWO TO ONE CONCLUSIVE PROOF THAT GEIGER HAD BEEN ROBBED OF HIS RIGHTS

The *Miner* was hysterical in its delight. In a minor heading it asserted that "Stanton, Myers, Clark and Norris made a desperate attempt to delay the righting of an outrage upon the ballot box (in the election of Whiteside), but the honest senators, Democrats, Populists and Republicans, would not consent that the loathsome and detested bribe solicitor should remain in their presence a moment longer than necessary, and Whiteside was sent forth to wallow in his own filth, the latest example of 'purity in politics,' an object of scorn and contempt of all who knew him."

"Helena, Jan. 26.—(Special to the *Miner*)—Fred Whiteside was fired, 'bag and conspiracy,' from the Senate this afternoon by his fellow-Democrats, and, amid the silence of contempt and scorn, the arch-conspirator departed from the senate chamber, an outcast from society, dishonored and disgraced, and henceforth to become a pensioner of the Anaconda company."

The editor of the *Miner* was warming up. Another subdivision of the story, under the line, "Filthy Remarks of Whiteside," followed with this example of Butte's journalism as of 1899: "Personal representatives of Marcus Daly shouted with glee when the scoundrel assailed the virtue of the women of Helena and stated that any man who voted for W. A. Clark would pimp for his own wife and daughter. Desperate attempt of a perjured and conscienceless villain to stain the character of reputable men and women of Montana. Fred Whiteside, retiring senator from Flathead County made a carefully prepared half hour harangue. The speech was such as could have emanated from no one but a masculine strumpet, a political bawd, a well paid harlot of a conscienceless gang of party wreckers, whose false lights gleam on every headland of democracy in the State of Montana."

On the next day the grand jury reported that there was not sufficient evidence of bribery on which to base an indictment of William A. Clark. That action was taken in face of the fact that forty-four witnesses had been presented before it by the Attorney-General. The *Miner's* headlines on the happy event announced:

HIS VINDICATION IS COMPLETE

MR. CLARK'S SUPPORTERS FULLY EXONERATED
BY GRAND JURY
TESTIMONY WAS FALSE
DAMNABLE PLOT OF CONSPIRATORS CAME TO NAUGHT

In contrast Marcus Daly's *Anaconda Standard* headlined:

THEY SIMPLY FELL DOWN FLAT!

HELENA'S GRAND JURORS DIDN'T HAVE PLUCK TO FIND
ANYBODY GUILTY—ATTORNEY GENERAL OF MONTANA
GIVES THEM SCORING IN RED HOT SPEECH

And thereupon Clark's vote, which had climbed to thirty-six during the two weeks of the grand jury and Whiteside hearings, gained four more. On January twenty-seventh it reached a total of forty, while Conrad's original thirty-eight slipped to twenty-nine. Despite the "vindication" by the grand jury, gossip of bribery increased rather than diminished with the increase in the vote for Clark. Three more votes slipped into the Clark column, the next day.

The joint session of the legislature buzzed with new rumors. A second vote in the one day, something unprecedented, was called for. Word was rife on the floor, in the galleries and in the lobbies, that everything was fixed. And on the next and final vote eleven Republicans abandoned their candidate and cast their votes for William Andrews Clark. The *Miner* headlined to the world:

VOICE OF THE PEOPLE HEARD

TRIUMPHANT ELECTION OF W. A. CLARK TO THE UNITED STATES SENATE

The story reported that Clark had been elected by a total of fifty-four votes out of ninety-three cast. It explained the suddenly acquired Republican support as having been decided upon by a caucus of the Republican members, basing their approval upon Clark's pledge that he would favor a protective tariff upon copper, wool and a few other items which made up the state's chief sources of revenue.

The *Standard* announced:

BRIBERY TRIUMPHANT IN REPUBLICAN PARTY

SCORN AND LOATHING ALREADY THE WAGE OF THE ELEVEN WHO VIOLATED PARTY OATH IN CASTING THEIR VOTES FOR W. A. CLARK—EXECRATED BY THEIR OWN PEOPLE—SHUNNED BY HONEST MEN THEY TOOK THE ARCH-BOODLER'S GOLD

The *Standard* minced no words and hesitated at no fear of a libel suit. Incidentally it was not sued, although it commented upon the newly elected Senator in this language: "He wanted to be elected by Democratic votes if possible, but he could not get more than a total of 38 Democrats to save his black soul. . . . Late Friday night, John B. Wellcome, the manager of Clark's boodle fund, was closeted with several of the recreant Republicans and fixed up the deal. It had to be made, else the Clark boom would have collapsed as sure as fate. All the week the Clark boodlers have been trying to buy up the requisite number of Democratic votes. Conrad and Maginnis men have been offered during the last few days all the way from $20,000 to $35,000. The boodlers cannot deny it. As a matter of fact, they do not deny it.

". . . All told Clark's election must have cost him fully $750,000."

But the tumult and the shouting did not die. Far from it.

CHAPTER XIII

CLARK'S "VINDICATION" AND DALY'S VENGEANCE

MARCUS DALY and his cohorts had been knocked as flat as they accused the grand jurors of falling. But they were up at the count of nine, perhaps a little groggy but still in the ring. Neither Daly nor his *Anaconda Standard* had any intention of suing Clark or the *Butte Miner* on charges of libel based upon the conspiracy charges. That was a sword which might cut two ways. Daly had too many millions of dollars to need another from the Clark fortune, even if he could have collected it. In contrast to some of the great money-makers of his time it was not his practice always to wring the last dollar out of every deal. Also he was aware that even a verdict of a million or so as punitive damages would not destroy Clark. What he wanted was to keep Clark out of the United States Senate. That, he believed, would hurt.

So, basing his action upon the fact that Clark had publicly accused him of conspiracy, he set out to prove the opposite. Henry C. Stiff, Speaker of the Montana House, A. J. Campbell, Congressman from Montana, Fred Whiteside, and several members of the State Democratic Central Committee, gathered with Mr. Daly to plan the continuation of their battle. Daly agreed to finance the investigation and the presentation of evidence before the United States Senate Committee on Privileges and Elections. This meant paying for counsel, obtaining the attendance of witnesses before the committee, and so forth. For this purpose Daly immediately donated twenty thousand dollars, which was put into the hands of Mr. Campbell as counsel for the informal committee. Later Daly added a few thousand more, but though no limit was put upon his pledge the total amount used was less than twenty-five thousand dollars. Subsequently he attested that fact under oath before the Senate committee.

The Daly committee immediately set to work. A memorial was drawn up for presentation to the Senate, charging that Clark had been elected to the Senate by bribery, fraud and other corrupt practices, and that a seat in that body therefore should be denied him. The names of twenty-five members of the Montana legislature were signed to that document. As soon as the Daly committee's inquiry had advanced sufficiently a second and similar memorial was drawn and signed by Robert B. Smith, Governor of Montana; T. E. Collins, State Treasurer; Henry C. Stiff, Speaker of the House; Charles S. Hartman, erstwhile Representative in Congress; and Congressman Campbell.

While this material was being prepared the interests opposed to Clark's ambitions were striking at various points in the Clark machine. Fred Whiteside, who had brought the original bribery charges and the thirty thousand dollars in currency evidence on the floor of the Assembly, decided that he had been libeled by Clark's *Miner* when it called him "a moral leper, a political bawd, a perjured and conscienceless villain, a masculine strumpet," and so forth. Unlike Daly, Whiteside felt that he could use even Clark money if it could be obtained on a verdict for punitive damages. He brought suit against the Miner Publishing Company, W. A. Clark, Alex J. Johnston, A. H. Wethey and Harry Sultzer as directors of that company.

The charges made under oath in that libel action, together with the deposition of William A. Clark taken in answer, were attached to the memorial to the Senate as corroborative evidence. The summary of testimony before the grand jury which had declined to indict Clark for bribery was attached as another exhibit. But more than that was needed by the protesting forces.

An action was instituted before the Supreme Court of Montana for the disbarment of John B. Wellcome, on charges of conduct unbecoming an attorney practicing before the bar of that virtuous state. Wellcome, it will be remembered, was the man named by Whiteside as having been the Clark agent who produced the thirty thousand dollars in currency which was displayed to the legislature as evidence of bribery.

But the Supreme Court decided that it had no jurisdiction in the disbarment proceedings. That did not dismay the opposition to Clark. They amended the charges in such manner that the court altered its first decision and agreed to hear the evidence. This evidence, having thus been recorded before the Supreme Court, was then in form to be attached to the memorials to the Senate as another exhibit in substantiation of the charges against Clark.

Even more satisfactory to the Daly contingent, new evidence was recorded in that hearing concerning the eagerness and unscrupulousness of the Clark cohorts in the offer of bribes. Dr. William Treacy, a Republican but a staunch supporter of Clark, approached Mr. Justice William H. Hunt, one of the three members of the Supreme Court, with a proposition that Judge Hunt could have one hundred thousand dollars if the court would dismiss the action against Wellcome. Treacy, who was Judge Hunt's family physician, advised that the justice accept the offer.

This offer, Doctor Treacy later asserted under oath, was made merely to test his friend's honesty. When cross-examined before the Senate committee concerning it, he insisted that he had never been authorized by the Clark interests or any other interests to make the offer, and that he had never had one hundred thousand dollars or any large part of it to carry through the test. Questioned as to whether it was the part of an honorable friend so to test the honesty of another friend in public office, his dignity was sadly battered.

Doctor Treacy evidently had a penchant for testing the honor of public officials. His offer being declined by Judge Hunt, though apparently with no great indignation on the part of the jurist, Treacy approached Attorney-General Nolan, in charge of the proceedings against Wellcome. He offered Nolan one hundred thousand dollars to drop the proceedings against Clark's counsel. This offer he repeated three times with no more success than in Judge Hunt's case, but with the adding of valuable ammunition to the supporting evidence against the Clark crowd.

Although no evidence was adduced that the Clark interests had authorized the offer of a bribe of one hundred thousand dollars to either Judge Hunt or Attorney-General Nolan, both documentary and verbal evidence was presented to show that Senator-elect Clark was taking a very warm personal interest in his counsel's battle against disbarment.

When that evidence was in there remained no shadow of a doubt that Clark, through an agent, had made an entirely unethical and improper approach to Chief Justice Brantley of the Supreme Court in behalf of his friend and counsel who was charged with having given bribes in his behalf. No less respected a man than the Reverend A. B. Martin, president of a girls' school at Great Falls, was Clark's go-between in this effort to influence the court. Both Clark and Brantley were trustees of the school. Under date of October 7, 1899, while the Wellcome disbarment proceedings were still pending, Clark wrote to the Reverend Mr. Martin urging him to see the chief justice and endeavor to prevent the disbarment of "that splendid man," John B. Wellcome.

The reverend gentleman, later insisting that he had no thought but to do a kindly deed, carried the Clark letter to Justice Brantley. The justice looked it over and returned it. Mr. Martin returned to his girls' school with the consciousness of a duty well done in behalf of the richest man in Montana and a trustee of the school which paid his salary. But the letter was preserved, to be discovered by the anti-Clark forces and added to the ammunition against William A. Clark in the hearing before the Senate committee.

John B. Wellcome was disbarred.

The outlook for William A. Clark was growing dark. The two memorials, signed by legislators, governor, treasurer and other officials of Montana, and backed by various documents and a list of nearly ninety potential witnesses, were presented to the Senate Committee on Privileges and Elections on December 4, 1899, eleven months after Clark's election, and three months before the beginning of the term to which he had been elected. The Senate had listened to an outline of charges and supporting evidence and directed its committee to investigate.

On January fifth, witnesses subpoenaed from Montana arrived in Washington, and the hearings opened. The committee was made up of Senators Chandler (chairman), Hoar, Pritchard, McComas, Caffery, Pettus and Harris. Charles J. Faulkner and Roger Foster appeared as counsel for Clark. George F. Edmunds, J. M. Wilson, A. J. Campbell and Charles S. Hartman appeared as counsel for the memorialists demanding a refusal of the Senate seat to Clark.

The hearings continued almost daily for more than three months. Nearly one million five hundred thousand words of testimony and argument and documentary evidence were recorded. Printed, they filled twenty-seven hundred and two pages in three heavy volumes of government reports. At the conclusion of the hearing the committee noted as "admitted or undisputed facts," the following:

> Prior to 1895, the elections in Montana were accompanied by enormous expenditures of money, unquestionably involving widespread belief that extensive corruption was resorted to in all elections. Ex-Governor Hauser, a witness called by Clark in his defense, testified that the expenditures made in connection with the contest to determine whether Anaconda or Helena should be the capital of the State were upwards of $1,000,000. Senator Clark admitted that he spent over $100,000 in that contest.
>
> Clark was a candidate for Congress in 1888 and was defeated. In 1890 there were two legislatures in Montana. Clark was elected United States Senator by the Democratic legislature but was not seated. In 1893 he was again a candidate but because of a deadlock there was no election. In 1895 he was the Democratic candidate but Senator Carter was elected. In the summer of 1898 the Clark interests organized a committee to secure the legislature which would meet in January, 1899.
>
> This committee consisted of his son, Charles W. Clark, William McDermott and A. J. Davidson. Charles W. Clark was made treasurer. To this committee Senator Clark gave unlimited authority to expend money which he agreed to furnish; an estimate, however, being made that at least $35,000 would be necessary to secure the State convention and that $75,000 might be necessary to secure the State legislature.
>
> Payments by Senator Clark to this committee and agents, admitted by Senator Clark, amounted to $139,000. In addition he gave $5,000 each to McDermott, Wellcome, his attorney, and E. C. Day, a member of the legislature.

Neither Clark nor the members of the committee made the sworn returns required by law.

In December, 1898, Clark closed a deal with H. W. McLaughlin, a member of the legislature, to buy lumber properties for $24,684, an amount described elsewhere in the report as "extravagant." McLaughlin voted for Clark.

E. P. Woods, a member of the legislature who had had to put a second mortgage on his ranch property in November, 1898, was able to pay it off in April, 1899. Woods voted for Clark.

W. E. Tierney was senator and W. C. Eversole and Michael Shovelin were representatives from Broadwater County. They operated three hardware stores and were heavily in debt when the legislature convened. After the legislature adjourned, March 3rd, Shovelin deposited $8,500 in the Montana National Bank. Eversole, who had been only a clerk, bought out Tierney's interest for $4,600 of which $1,500 was paid in cash. Tierney was able to organize the State Bank of Townsend and subscribe $5,800 for capital stock, while Shovelin subscribed $1,000 and Eversole $1,000, all cash. All three voted for Clark.

D. G. Warner, a member of the legislature, in January, 1899, received $7,500 in bills from E. L. Whitmore, confidential agent of Charles W. Clark, for small parcels of real estate in and near Boulder. Warner voted for Clark.

H. H. Garr, who had to borrow $25 to go to Helena as a member of the legislature, managed to pay $3,500 in currency for a ranch after the session.

John H. Geiger, a man of no means, who was seated as senator after Whiteside's dismissal, although a Republican, voted for Clark. After the session he took home with him $3,600 of which he said he had made $2,500 gambling and found $1,100 in his room, evidently dropped through the transom.

Stephen Bywater, a Republican, employed as a railroad conductor, voted for Clark for the first time on the eighteenth ballot, on which Clark was elected. Immediately after adjournment he deposited $15,000 in the Montana National Bank. He explained by saying that he had brought $6,000 in currency from home, and had obtained $9,000 from his brother, but offered no corroborative evidence in support of that explanation.

W. W. Beasley, a Republican, who owed a $400 board bill at the time of his election, first voted for Clark on the eighteenth ballet. The next night he left Helena for St. Paul with $5,000 in currency in his possession.

All those details the Senate Committee's report characterized as "admitted or undisputed facts." Concerning the Whiteside charges that John B. Wellcome, as agent for Clark, had given thirty thousand dollars in currency for the votes of Whiteside, W. A. Clark of Virginia City, Myers and Garr, the committee noted the denials of Wellcome and other witnesses, but added that "the majority of the Senate Committee finds these bribery stories in general true despite the denials of Mr. Wellcome."

"In the opinion of the majority," the committee's report continues, "there was no affirmative disproof produced by Senator Clark at any time showing that the facts in connection with the $30,000 exposure were not true. The charge that the accusations and the exposure resulted from a conspiracy formed and carried forward by Mr. Marcus Daly was a necessity of Mr. Clark's canvass and defense. . . .

"The reasons given for the Republican votes for Mr. Clark seem to be pretenses and covers. . . .

"In 1898 when Senator Clark renewed his canvass for the United States Senate, and his friends determined to directly disregard the law, he had no right to spend more than $1,000 through his own committee, nor more than $1,000 for his personal expenses. He expended a sum vastly in excess of the limit provided by law and failed to render any return; and no return was made by the aforesaid persons acting in his behalf. This failure to make return was with the full knowledge of the fact that it was widely charged and widely believed that the election had been accomplished by bribery and the corrupt expenditure of money on a very large scale.

"He also, in June, 1899, with full knowledge of this belief and these charges, but in accordance with what he testified was his custom, destroyed all his checks, including those for his expenditures in his canvass, and was unable to give any full and satisfactory vouchers for the money so expended by his agents.

"The committee find the case proved without relying upon such destruction of checks, failure to render accounts, and disregard of the limit of legal expenditure. But they furnish to the committee strong additional reasons for believing the other evidence, which seems to them to establish the corruption charged. . . .

"It is also a reasonable conclusion upon the whole case that Senator Clark was fairly to be charged with knowledge of the acts done in his behalf by his committee and his agents conducting his canvass. He arrived in Helena from Butte on January 4 and remained there until after his election on January 28, and was in constant conference with his committee and his agents."

All that having been decided, "The finding of the committee is that the election to the Senate of William A. Clark, of Montana, is null and void on account of briberies, attempted briberies, and corrupt practices of his agents, and of violations of the laws of Montana defining and punishing crimes against the elective franchise."

Clark had lost that round in his long and bitter battle with Marcus Daly. He was on the ropes, but he was able to see the hand of the time-keeper reaching for the bell before he could be counted out. He still had his millions, his ambition and his determination. He was fighting with his brains, and they had not been addled by the knockdown. Brainwork and footwork were still coordinating.

Before the committee's report could be presented to the Senate, and before that august body could declare the Clark seat vacant, he resigned. But that was not all. Far from it. Before he resigned he had prepared the way to foil both the Senate Committee and the Senate itself.

Governor Robert B. Smith, one of the Montana officials who had signed the memorial to the Senate charging Clark with bribery and asking that his election be declared null and void, had been inveigled out of the state on a trumped-up business call. The Lieutenant-Governor was also out of the state at the same time. But like a well-trained shortstop, the Lieutenant-Governor had his eye on second base, and as soon as that sack was uncovered when the Governor faded back toward right field to catch a fly, the Lieutenant-Governor took his place. He was just in time to take the throw from Clark in the form of Clark's resignation, and he immediately appointed Clark to the vacant seat in the Senate.

Governor Smith and all the Marcus Daly cohorts in Montana were incoherent with rage. Clark's *Butte Miner* and a dozen smaller Montana newspapers controlled by the Clark interests announced blatantly that Clark's "vindication" was complete. Governor Smith did not feel that way about it. He returned to Helena posthaste, rescinded the appointment and announced that it had been made without authority, through trickery as fundamentally corrupt as Clark's original election.

Clark, being a wise man, accepted the "vindication" for what it was worth, or perhaps a little bit more. He did not attempt to take the seat to which he had been appointed. Instead, he started a campaign for reelection at the next session of the legislature. That campaign began in the autumn of 1900.

The *Anaconda Standard* renewed its fight in characteristic fashion and outlined its opposition campaign policy by announcing "the receipt of the following new books, reviews of which will appear later: 'The Theory and Practice of Boodling,' by Prof. John B. Wellcome, ex-Principal of the W. A. Clark Boodling Institute, author of 'A Guide to the Barrel;' 'The Boodler's Pocket Companion,' 'Recent Progress in the Art of Boodling,' etc., etc.; 'Reminiscences of the United States Senate, or The Four Hottest Months of My Life,' by Wm. Andrews Clark, Doctor of Boodle, author of 'An Star Spangled Banner,' 'The Rise and Progress of the Barrel in the 19th Century,' 'My Untarnished Reputation,' a fairy story."

In the meantime, however, the burning brand of Standard Oil had been put upon Marcus Daly, the Anaconda company and its holdings by the organization of a holding company announced to the world as the Amalgamated Copper Company. This branding was revealed as Clark's new campaign gained momentum. In October, 1900, the *Northwest Tribune.* spoke its piece:

"Who are bringing these charges of bribery against W. A. Clark? His political enemies, and the corporations that wish to steal his property and the political rights of the vast majority of the people of Montana.

"How have these corporations gone to work to get control of the state? By buying Lee Mantle and his newspaper. . . . By establishing a kerosene sheet here in Helena, and by buying a political fakir lately indebted to Senator Clark for a government job for which he was never fitted, to edit that paper. That looks like bribery.

"In addition these corporations have acquired as political assets John R. Toole, former chief engineer of the Anaconda Copper Mining Company's slush fund; A. J. Campbell, discredited and disgraced by his own acts; and the great 'friend of labor,' Thomas S. Hogan, who one week before the meeting of his own convention in this city said he would do anything to prevent this state from falling into the hands of the Standard Oil and Amalgamated Copper trusts; tempting him to sell every principle that he ever stood for. That looks like bribery, but not Clark's bribery. . . .

"While the Democrats propose to send Clark to the Senate, the Standard Oil and Amalgamated Copper trusts demand the reelection of Thomas H. Carter. It is well known that thirteen checks for $1,000 each, signed 'Thomas H. Carter,' bought his election; it is well known that these checks were repudiated. Whatever else may be said of Clark, no one can say he ever repudiated a debt. Carter's election was won by bribery, but not by Clark's bribery."

The fight was warming up again. A man who could build up a fortune approaching fifty million dollars, and that without any stock-market business whatsoever, who could give personal management to a wider variety of business interests than any man of his day in the United States, who could admit spending one hundred and fifty-four thousand dollars on a personal political campaign when the law limited him to two thousand dollars and then claim "vindication" through an unauthorized appointment to the United States Senate which had declared his election void because of corrupt practices, was not a man to be ignored.

The new campaign in Montana was a wild and gay one. The miners and rustics were entertained as miners and rustics have never been entertained since. Newspapers throughout the country found thrilling copy in the details of the campaign. Montana politics had become a subject of national interest. Witness extracts from an article published by a metropolitan newspaper in the autumn of 1900:

"Helena, Oct. 24.—W. A. Clark himself is engaged in a tour of the state seeking vindication, as he terms it. He is accompanied by his band, whose thirty-five members receive $6 a day and all expenses; the Charlie Clark drum corps of twenty-five members at $5 each, and a dozen or so imported spellbinders. A carload of fireworks is sent out each week from headquarters to make a proper display. For example, in Helena last Saturday night some $2,500 worth of red fire and fireworks were distributed among the urchins of the town to welcome Clark, although this is the only town in the state where he is received with any degree of enthusiasm. In addition the men got $10 each to carry transparencies, picks, tools, etc., to show the esteem in which the laboring element holds him.

"Daly is not much of an orator and besides his health will not permit him to canvass the state himself, but he is represented by some of the ablest politicians in the West who are making Clark's pathway anything but pleasant, picturing him as a convicted briber and corruptionist. Perhaps the most persistent man on Clark's trail is Governor R. B. Smith, who was inveigled out of the state in order that the Lieutenant-Governor might reappoint Clark on the day his resignation was tendered to the acting Governor.

". . . Both men sent agents all over the state purchasing country newspapers. Sometimes the rivalry was so keen that the country editors got four or five times the original costs of their plants, together with a year's contract as managing editor . . . with the result that each county seat has its Clark and its Daly paper and the wire service of the telegraph company is inadequate to handle the daily report.

"Even the baseball league was not exempt from the influence of the feud. During the middle of the season, Helena and Anaconda were struggling for the leadership with a series between them in progress at Anaconda. Clark opposed Anaconda in the state capital contest and he naturally would favor Helena in the baseball struggle. So at a conference of Democrats in Butte he proposed to take the whole party to Anaconda to root for Helena. The thought was no sooner conceived than put into execution, and purchasing several thousand yards of red and yellow ribbon, the Helena colors, which were cut up and pinned on the delegates, he chartered a special train and started for Anaconda. The score was tied but the rooters soon had the Anaconda club in the air, and Helena had the lead.

"But Daly was not to be outdone. Upon the arrival of the Clark special at the fairgrounds, the friends of Daly sent emissaries up town to search for small boys who were armed with bells, horns and tin cans. The boys were admitted free and certainly made an unearthly noise but they came too late and Helena won out. Clark is credited with saying that the victory was worth $5,000.

"Then again, Clark has a vaudeville manager in Col. Wm. Thompson, lately manager of the Boston Lyrics, who is said to receive a large salary for providing the speakers with actors to draw crowds whom Clark orators may then address, and who would not otherwise assemble. Thompson failed when he tried to get James Neill to spring Clark jokes for $1,000. He tried three times before he realized his blunder. . . . There are many ludicrous phases in Montana politics.

"It is this turmoil which gives hope for Republican success at the polls. The state was never so prosperous as under the present administration. More miners are employed and at better wages; cattlemen are receiving more for their steers; sheep-men get from 17 to 20 cents for their wool against 7 cents under Cleveland. Yet many of the voters seem wedded to the 16 to 1 idea as the only panacea for ills that do not exist . . . due to the fact that a majority of the laboring class are foreign and ignorant."

Civilization in Montana in 1900! Gay days! Red fire! Free vaudeville! Marching brass bands! Torchlight processions! Bribery taken seriously only by those who profited or those defeated! Prosperity! Excitement!

Marcus Daly's grip had weakened. Suffering from an incurable illness, he was unable to direct in person his campaign of opposition to Clark. Having turned his holdings in to the Amalgamated Copper Company in the previous year and so being branded with the Standard Oil mark, he had sacrificed much of his personal popularity. The Clark interests, ably assisted by F. Augustus Heinze, who had developed into a soapbox orator of power, had smeared the opposition with oil and scented it with kerosene. The ignorant majority of Montana voters were as ready in 1900 to believe any demagogic outbursts against trusts and capitalism as voters of their character are today.

A large part of Clark's ticket went into the new legislature. Marcus Daly died in New York on November 12, 1900. Men and women may still be found in Montana who insist that he died of a broken heart. The able physicians who attended him, less sentimental, reported that the cause of death was Bright's disease, with complications. In any event, Clark's most implacable and powerful enemy had been removed.

Wise in his experience of two years earlier, Clark's supporters for the senatorship fought the final battle in caucus instead of upon the floor of the joint assembly. Ten Populists joined the Clark Democrats. That was enough. On January 16,1901, Clark was elected to the Senate by a vote of fifty-seven against Senator Thomas H. Carter's thirty-one, with four for Martin Maginnis and one for Robert B. Smith.

Now he could exclaim with greater pride that his "vindication" was complete. But such satisfaction as he might take in that must have been at least slightly impaired by the realization that Marcus Daly, his bitter enemy of more than twenty years, would never know of his victory.

And still the battles which turned the eyes of the nation upon Montana were not ended. F. Augustus Heinze had arisen to a more spectacular leadership than had Marcus Daly. A new phase of warfare was under way.

CHAPTER XIV

STANDARD OIL SHOWS THE WAY

WHILE Clark and Daly had been battling through the years for political control of Montana, either as a matter of personal enmity or for what it might give them in the way of legislation helpful to their interests, Butte had proved itself the "richest hill on earth," as Daly had prophesied. The metal output of the district each year was enough to load a freight train twenty miles long. Annual production of refined metals—chiefly copper, but including zinc, lead, silver and gold—was worth more than fifty million dollars.

That was big business—big enough to attract even the eye of Henry H. Rogers of the Standard Oil Company, then the most powerful if not the most feared and hated trust developed in the advancing civilization of the United States. A description of the eyes of Henry H. Rogers, written by Thomas W. Lawson who looked into them all too often, and published first in *Everybody's Magazine,* copyrighted by the Ridgway-Thayer Company, and in book form under the title of *Frenzied Finance,* in 1905, should be illuminating here:

"One must see Mr. Rogers' eyes in action and in repose to half appreciate their wonders. . . . I have seen them when they were so restfully blue that I would think they never could be anything but a part of the skies that come with the August and September afternoons. . . . I have seen them when they were so green that within their depths I was almost sure the fish were lazily resting in the shadows. . . . And I have seen them as black as that thunder-cloud which makes us wonder, 'Is He angry?' And then again I have watched them when they were of that fiery red and that glinting yellow which one sees only at night when the doors of a great, roaring furnace are opened.

"There is such a kindly good-will in those eyes when they are at rest that the man does not live who would not consider himself favored to be allowed to turn over to Henry H. Rogers his pocketbook without receiving a receipt. They are the eyes of the man you would name in your will to care for your wife's and children's welfare. When their animation is friendly one would rather watch their merry twinkle . . . than sit in a theatre or read an interesting book; but it is when they are active in war that the one privileged to observe them gets his real treat, always provided he can dodge the rain of blazing sparks and the withering hail of wrath that pours out on the offender. To watch them requires real nerve, for it is only the nimble, stouthearted, mail-covered individual that can sustain the encounter."

The eyes of Henry H. Rogers, then, had been attracted to the millions of dollars pouring from the mines of Butte. He was the chief executive of the Standard Oil trust. He knew and approved of the methods by which that organization had crushed competition and gathered to itself the greatest wealth and the greatest power concentrated in the hands of one small group of men since the fall of the Roman Empire. He knew that there were other ways of making a mine pay a profit than by mining and selling its metal. He knew the tricks of Wall Street, the watering of stocks, the organization of holding companies, the demoralization of public officials, the appeal to human greed, the exploitation of ignorance, the profits of monopoly.

Henry H. Rogers never overlooked an opportunity. He was never constrained by consistency in the middle-class ethics of life. He was the sort of man who could, and did, go straight from the church pew which he occupied with religious devotion, to a secret conference designed to win millions from men of lesser power but similar ambition. His whole economic and social philosophy could be summed up in his own words as quoted by Lawson.

"No man has done his business properly who has missed a single dollar he could have secured in the doing of it. I do not think a fair judge would find me guilty of avarice, either in business or in the manner of my living, and yet I am fairly miserable if I discover that, in any business I do, I have not extracted every dollar possible. It is one of the first principles Mr. Rockefeller taught me; it is one he had inculcated in every 'Standard Oil' man, until today it is a religion with us all."

Such a man was Henry H. Rogers of the Standard Oil school when J. Edward Addicks, of similar character but less strength and capital, expanded with the profits of his gas monopoly in Boston, decided to grab control of the gas business in Brooklyn. Standard Oil men controlled Brooklyn gas. Standard Oil men were not in the habit of allowing anyone to encroach upon their preserves. They promptly countered by entering the Addicks gas preserve in Boston. The contest raged for more than two years, in the stock markets, the courts and elsewhere with such violence that the Bay State gas war will always remain a classic of business warfare in America. Needless to say, the Standard Oil crowd won. Millions were the prize of that victory, but more important to this narrative, his contacts in Boston had given Henry H. Rogers his first thrilling vision of the possibilities of the copper industry.

Boston had been copper-conscious for more than half a century—ever since Louis Agassiz had announced that there was copper in large quantities in northern Michigan. It had been making a profit out of that copper since E. J. Hulbert had opened up the Calumet and Hecla ore body in 1864. It had started to develop what promised to be a monopoly of copper production in America when Quincy Adams Shaw, Horatio Bigelow and half a dozen other Back Bay aristocrats began to buy up prospects in the neighborhood of the Calumet at the close of the Civil War. By the time the Calumet and Hecla properties were consolidated in 1871, Boston residents had received nearly all of the two million eight hundred thousand dollars in dividends paid by those properties. Boston had seen Calumet and Hecla, owned almost entirely by its residents, advance in twenty-five years from one dollar to four hundred and fifty dollars a share.

Boston families had taken in some one hundred million dollars in dividends, mostly from Calumet and Hecla, partly from Butte. Copper millionaires were thick in Boston. Dozens of mansions had been built with copper dividends. The copper aristocracy had established itself as firmly as the earlier aristocracy which had grown rich through the indirect swapping of clipper ship cargoes of rum for cargoes of slaves dragged from African jungles.

Yes, Boston was copper-conscious. And Henry H. Rogers, battling with J. Edward Addicks for the millions involved in the Boston gas monopoly, was entertained in the homes of Boston's copper millionaires. When such tangible evidence of wealth surrounded him, a man of Rogers' character was quick to learn the source and history of that wealth. He also became copper-conscious. Very soon he learned a great deal about copper.

Rogers made himself acquainted with the fact that the Rothschilds, as great a power in the finances of Europe as was Standard Oil in the economics of America, had seen fit to buy the Hearst estate interests in Butte and Anaconda. He learned that Lewisohn Brothers, the leading copper sales organization in the world, had made millions through a virtual monopoly of the distribution of copper. In evidence of that, and in further evidence of Rogers' character in business, he should be quoted as his statement was set down in Lawson's *Frenzied Finance.*

"Let me show you, Lawson, how you have overlooked the best part of the copper business. We have found that for years Lewisohn Brothers have had a double-clamped and riveted contract with half the best producing mines in the country to sell their output, and they have grown very wealthy. As near as we can make it, they have made at least fifty million dollars in one way or another in the last ten or twelve years.

"First they have had a big profit as their commission for selling; next, big interest out of the advances they make to companies while their output is being sold; now, they actually control the copper market of the world. Think of it, Lawson, for a few seconds and the possibilities will loom up to you. You can buy or sell any number of millions of pounds in futures or actual deliveries. Suppose a man controlling the selling of three or four hundred million pounds a year should knock the price to, say, ten cents, sell to himself the year's output of all the mines he controls, and then lift the price to, say, twenty cents. He would have a sure profit with absolutely no risk, of thirty to forty millions of dollars. Wouldn't he?

"Then if, before he broke the price, he sold copper mining stocks short, and if, before advancing the price, he covered and loaded up on them, he could easily make an additional thirty or forty millions. Think it over, and you will agree with me that the possibilities are far beyond those of oil, and perhaps at the same time you can account for the violent fluctuations in copper stocks and the price of the metal during recent years.

"A man in such position could absolutely dictate to all new mines whose selling agency he could secure under long-term contracts. When their stocks were up, he could pinch them to the edge of bankruptcy by refusing to sell their metal or advance them the cash they needed for operation. . . .

"We have dug up some queer doings during our investigation and I think I can put my finger on a great many millions of dollars now in the hands of certain mine officers which could be recovered by the different companies they have been acting as trustees of. It would be quite an eye-opener to some of your pious Bostonians to know that the controlling officials of several mines are silent partners in some of the big selling agencies. . . .

"To come back, Lawson, to Lewisohn Brothers. We must buy that concern. . . . . Leonard Lewisohn, head of the firm, is second to no man in America as a business man, which means he will not hanker for a fight with us; and when I show him we will buy, if necessary, the control of all the companies they represent, he will see the absolute futility of opposing us. . . .

It must be clear that Henry H. Rogers, of Standard Oil, had all the essential information, business qualities and experience to see in copper a greater "trust" than the world had ever known. "Think it over, and you will agree with me that the possibilities are far beyond those of oil."

The foundation was laid for a new organization in Butte—a step onward in the advance of "civilization" as revealed at the end of the nineteenth century, twenty years after Butte itself had become copper-conscious and started to prove itself "the richest hill on earth."

Butte doubtless had become a little snooty with its accumulated riches. Magnificent homes, imported furnishings, blooded race horses, vintage wines, had taken the place of log cabins, wooden stools, work mules and trade whisky in the lives of its leading citizens in twenty years. It was able to turn up its nose at its own rude beginning. But it had a sense of humor. A story, probably apocryphal, but nevertheless revealing, may be heard in Butte:

For a popular function which would now be termed a pageant, but was known in the 'nineties as tableaux, three young bucks of the Flathead tribe, complete with bows and arrows, had been brought to town to represent the primitive background of Butte society. The bucks were quartered in a bedroom of one of the new hotels of which the city was proud. When the program manager sent a messenger for them, only two Indians were in the room. Asked for their companion, they pointed stolidly to the bathroom. There the messenger found the third Indian, dead, with an arrow through his heart. There was an immediate alarm and a prompt investigation. The surviving Indians waved away the inquisitors and explained the incident to their own complete satisfaction with a phrase: "Him spoilum spring."

With such evidence of the primitive and the modern side by side Butte could be expected to do almost anything. It could see its richest citizen denied a seat in the United States Senate because of bribery and could help to return that same citizen to the Senate two years later. It could gather, almost in adoration, around the soap-box occupied by F. Augustus Heinze while that young man announced that there was a bill before the legislature which properly should be entitled "A Bill to Turn the Government of the State of Montana over to the Standard Oil Company," and it could then put pressure on its assemblymen to pass that bill. It could watch with the indifference of ignorance while Marcus Daly dickered to turn the "richest hill on earth" over to "the greatest trust on earth." Then it could be lashed into a frenzy by a biased report of what had happened. In one moment it could raise a colossal bronze statue of Marcus Daly in its streets, and in the next could so befoul and desecrate that statue that a great electric lamp had to be raised to protect the monument from defilement by lesser men.

Clark was busily engaged in the campaign to destroy Daly's power in Montana by electing a legislature which would elect him improperly to the Senate when Daly began his negotiations with Henry H. Rogers in 1898. Heinze was up to his ears in the task of making good his threat to give A. S. Bigelow of the Boston companies "a fight that will be heard of from one end of this continent to the other." Daly, already broken in health, was seeking to consolidate his wealth under the protection of such men as Rogers and William Rockefeller.

Clark, as he had been from the beginning, was an independent operator, having nothing to do with any business which he did not control. But he was not too independent to see the advantage of enlisting the support of the popular Heinze for political purposes. If, in so doing, he tacitly gave aid and comfort to Heinze in the apex suits against the Boston companies it would cost him nothing, and might embarrass his chief enemy, Daly. Clark's own mines had not been threatened by Heinze. The mines of Clark's enemies were threatened. In the election of Judge Clancy and in minor political contests in Butte Heinze had proved himself able to swing seventy-five percent of the local vote.

Heinze was as popular with the muck-smeared miners whom he met and treated in the saloons upon the hill as he was with the elite whom he entertained at exclusive dinners under careful chaperonage. He was the perfect complement to the cold and somewhat supercilious Clark. He was a man for Clark to tie to—until Clark's purposes had been accomplished. Clark, with many millions available, appeared to Heinze to be a potentially valuable ally.

That, then, was the line-up: Clark, with the support of Heinze, against the Daly crowd in politics; Heinze against the Boston interests in mining; Clark for himself alone in business, but giving tacit approval to Heinze's fight to grab the Boston properties before the "trust" could get them; Daly, weakened by illness, preparing to sell out to the Standard Oil crowd.

In promotion of his own interests when the state campaign of 1898 had been lined up to his satisfaction, Daly went to Henry H. Rogers. Rogers, as explained above, had recently learned a lot about the potential profits of copper. He was in an expansive and receptive mood.

Daly found Rogers most cordial. No one in America knew better than Henry H. Rogers how to be cordial, or how to pump a business caller. He exerted all his charm upon the Irish miner who had become a copper magnate. After the meeting he called in Thomas W. Lawson, Boston broker and promoter specializing in copper stocks, who had been doing preliminary work for the Standard Oil group with reference to entering the copper market. Rogers told Lawson that Daly had made an offer of seventy-five million dollars' worth of property for thirty-nine million dollars.

According to Rogers, as quoted by Lawson, Daly had been concealing from his partners, Haggin and Tevis, and the Rothschilds who had bought the Hearst interests several years earlier, the true value of the Butte mines under his management. He had been taking out millions enough to satisfy the owners, but leaving more millions easily available against the time when he might be able to buy out his partners and secure the treasure to himself. Whether that story is true, or whether Daly shrewdly allowed himself to be "pumped" by the complacent Rogers in order to put through the deal, must remain a question. All the principals are long since dead.

In justice to Daly, however, it should be said that outside of that unsubstantiated charge of perfidy there is nothing in the available record to indicate that Daly was ever a man to doublecross his friends. It is known that Haggin, with the approval of Tevis, cooperated with Daly in the subsequent negotiations with Rogers. It seems probable that Daly, who had proved himself an inspired geologist, a practical mining engineer and a competent political and business organizer in the development of Butte and Anaconda, might have told Rogers there were far more millions available in the hill than had been taken out. His faith in Butte had always been a faith literally to move mountains.

In any event, the deal was consummated. The Amalgamated Copper Company was organized as a holding company to take over the control of the Anaconda properties, and the near-by Colorado, Washoe and Parrott, at a private valuation of thirty-nine million dollars—to the public seventy-five million dollars. The Standard Oily hand of Henry H. Rogers closed upon Butte with that decision. Civilization had advanced in twenty years to a point where two dollars could be made to grow where one had grown before.

Butte, in marked contrast to the Comstock Lode of Nevada, to the later gold and silver camps of Colorado and the camps of Goldfield and Tonopah, had never been a stock-conscious town. It had done most of its gambling on faro layouts, roulette wheels, poker tables, foot races, and an occasional flyer on the horse races when Marcus Daly's bangtails were running on the local track. Copper metal, not copper stocks, had been the source of its wealth. Its mines had been worked under ground, not in Wall Street.

But with the entry of the magic name of Standard Oil into Butte through the organization of the Amalgamated Copper

Company, all that was changed. Suddenly one dollar's worth of copper, hidden a thousand feet below the deepest shaft, could be capitalized in the form of two dollars' worth of stock, and actually sold for cash, ten years before it could be mined. So had the economics of civilization advanced from the days when the first hungry miner on Silver Bow Creek had had to wash a tenth of an ounce of gold from the sand and carry it to a trading post, and exchange it for beans and whisky before he could eat and drink.

Butte's eyes opened, and the eyes of the world opened with them. The town had thought well of itself and of its resources before this momentous development. Now it thought twice as well of itself. Two dollars for one! Let joy be unconfined!

CHAPTER XV

STANDARD OIL TAKES OVER

IT was all very simple when promoted by such experienced gentlemen as Henry H. Rogers and William Rockefeller, backed by the millions of dollars which Standard Oil methods had wrung from a rich and credulous country. Henry Rogers' eyes, doubtless green at the time, saw the possibilities of a trust greater than Standard Oil. Henry Rogers' brain, clicking like a perfect mechanism, well oiled, evolved a plan whereby he and his associates could buy the properties which Daly appraised at thirty-nine million dollars, without the expenditure of one dollar of their own money. At the same time they could control the trust and guide most of its profits into their own pockets.

A campaign of national scope—one of the first great campaigns of economic propaganda to be attempted in this country—was organized. Thomas W. Lawson appeared to be the logical man to prepare the public for the flotation of the Amalgamated Copper Company. Lawson had won riches and fame through his methods of publicity in the great Westinghouse-General Electric warfare in the early 'nineties. He had added to that fame in the Rogers-Addicks war over Bay State gas. Everyone in the United States who followed the stock market in 1898 knew Lawson. He had built himself up through several years of clever publicity as the honest friend of the people. He had developed the practice of putting his facts and theories concerning the market before the people in cleverly worded statements widely circulated in purchased space in the newspapers.

Lawson was a shrewd trader himself, and possessed a happy and original talent for winning the confidence of the public through these apparently straightforward statements. On numerous occasions he had thus publicly advised the purchase of certain stocks, and made money for those who followed his advice. He had even beaten Wall Street on some occasions, and his cry of Victory had gone forth to the world. He had a following which awaited his market advice eagerly, and acted upon it promptly.

Being a Boston broker, he had always been interested in copper, and according to his own claims had been the first to inspire Rogers to the formation of a copper trust. Lawson therefore became the publicity man in the task of preparing the public to snap up the Amalgamated stock as soon as it was offered for subscription.

The job was well done—sensationally done. By the time organization was completed potential investors throughout the United States, and even in Europe, were falling over each other in their eagerness to get into this great new company of which Marcus Daly was to be president, Henry H. Rogers, Vice president; William G. Rockefeller, secretary-treasurer; and James Stillman, J. P. Morgan and Governor Flower on the board of directors.

Lawson had not failed to announce that directorate to the waiting world. It was a world that had confidence in the moneymaking genius and power of Rogers, Rockefeller, Morgan and Stillman. It was a world acquainted with the ruthless success of Standard Oil. It was a world that believed the National City Bank, which was to receive the subscriptions, was the Standard Oil's bank. It was a world willing and eager to believe Lawson's alluring advertisements over his own signature, stating: "It is probably the first time in the history of public subscriptions that a stock is worth and can be sold for 50 to 75 percent more than the subscription price." It was a world willing to believe Lawson's unqualified statement that the company "will from the start and always after, pay 8 percent dividends annually." It was a world willing to believe Lawson's assertion that the companies included in the amalgamation were at that moment earning twelve to sixteen percent per annum.

The organization was completed, the public "educated," the ballyhoo perfected without one dollar of Rogers, Rockefeller, Stillman, Morgan or Standard Oil money having passed to Marcus Daly for purchase of the Anaconda and other properties which he had appraised at thirty-nine million dollars. The Amalgamated Copper Company was capitalized at seventy-five million dollars, divided into seven hundred and fifty thousand shares of a par value of one hundred dollars each. How it worked is best described in the newspapers serviced by the *Boston Financial News,* on the day after subscriptions closed:

$412,000,000 FOR AMALGAMATED COPPERS
$75,000,000 SUBSCRIBED MORE THAN FIVE TIMES OVER
FINANCIAL WORLD COPPER MAD

"Subscriptions of $412,000,000—the largest in any financial deal in the world's history—are reported by the *Boston Financial News* to have been received toward the Amalgamated company. . . .

"One week ago, says the report, it was announced that the Standard Oil magnates, Rogers, Rockefeller, and their associates, had begun their conquering march upon copperdom. . . . Never before since the first dollar of civilization was invented to take the place of the stone tokens of barbarism had such a thing been heard of—$75,000,000 of stock to be sold to investors at

$100 per share, and in one week after the birth of the corporation upon which it was based.

"During the entire week from all parts of the world have poured into the City National Bank applications, accompanied by checks for the first payment—one continuous stream of entreaties—for some of the shares of this great enterprise. Nothing in history tells of such a movement. . . . Immediately after the subscription closed, 140, or forty percent premium, was bid for the stock. . . ."

Obviously it was then easy for the promoters to pay, in cash and stock, to Marcus Daly and his associates the thirty-nine million dollars at which they had appraised their holdings, and that without putting up any of the money of the promoters. Two dollars had been made to grow where one grew before, and the extra, unearned dollar, had been used to pay for the seed.

But even that was not all. The amount of subscriptions had been so overwhelming that even Henry H. Rogers and William Rockefeller, accustomed as they were to playing with millions, were astonished. But not for long. So many millions in hand immediately inspired a thought of how to get more. With the able assistance of James Stillman they rejected bids on various technical and legal grounds until the acceptable subscriptions were reduced to approximately one hundred and thirty million. Obviously they could not issue one million three hundred thousand shares of stock in a company having only seven hundred and fifty thousand capital shares. That would have been illegal. They never did anything illegal. They retained the smartest lawyers in the country to protect them from that.

So an arbitrary allotment of one share for each five sought was decided upon. This gave to the promoters some twenty-six million dollars cash, and left them some five hundred thousand shares in the company. In other words, they had obtained two-thirds of the specified cost from the public, and given that public only one-third of the stock. They retained the two-thirds, valued on the day after the books closed, at approximately seventy million dollars.

That was the high point of civilization to which Wall Street had advanced in 1899—a civilization revealed and proved in the character and success of such men as Henry H. Rogers and William Rockefeller. For the purposes of this narrative it affords an illuminating contrast to a day only one generation earlier when men were digging with elk horns in the first mine of Butte, upon which this astonishing achievement was based.

No longer were the mines of Butte to be operated solely for their production of copper, zinc, silver, lead and gold. Thenceforth they were to be operated also for their production of new stock certificates, theoretically based upon metal resources and output, but practically based upon the acquisitive genius of Henry H. Rogers, William Rockefeller, James Stillman, and their associates and successors in Wall Street.

With the astonishing success of the initial flotation of the Amalgamated Copper Company, Rogers and Rockefeller were prepared to grab additional millions through their control of the market and their knowledge of market practices. While scores of favored "friends" of the Standard Oil magnates, and thousands of minor investors were being induced to hold their stock and to buy more on the ground that it was worth far more than they paid for it, the principals began cautiously to sell. When the inner group had thus cashed their profits, the market was allowed to break.

The National City Bank called its loans upon the stockloans up to ninety percent of the par or flotation value, which it had made most cordially to make the original purchases easy.

The market for Amalgamated crashed. Marcus Daly died. The price of the stock tobogganed to seventy-five dollars, a loss of twenty-five per cent instead of the promised gain of fifty per cent. The suckers were squeezed out.

The power and ruthlessness of the organization may be revealed by the size of some of the suckers. They included the political boss of Brooklyn, "friend" of Standard Oil, down on the original subscription for six hundred thousand dollars; James H. Eckels, then President of the Commercial Bank of Chicago and Comptroller of the Currency under President Cleveland, subscribing for one million dollars' worth of the original stock. There was even J. Edward Addicks himself, of Bay State gas, subscriber for twenty thousand shares at two million dollars.

How many hung on while the price plunged from one hundred to ninety, to eighty, to seventy-five, is not in the record. Believing, as most of them undoubtedly did, the financial omnipotence of Standard Oil, probably as many held as could hold. But inasmuch as most of the larger suckers had hypothecated the stock at a large fraction of its purchase price in order to pay for it, many of them were unable to hold. As their security was thrown on the market by the lending banks and brokers, the rout was almost complete.

Then, when the price was down to a point where efficient management of the mines and smelters, and honest distribution of their profits could pay an earned dividend of eight to sixteen percent, the insiders started their second coup. They could buy at seventy-five, stocks which they had sold above one hundred, and "shorted" all the way down. With the millions in profits so obtained, they began to buy.

Under the adroit management of that genius of market manipulation, James R. Keene, the price of Amalgamated was driven up as steadily as it had been driven down, until it was again above the par at which it had been floated. An additional ten or fifteen million dollars had been added to the profits of the Standard Oil group, without the slightest change in the intrinsic value of the mines and smelters of Butte and Anaconda upon which the stock was based.

Everything was ready for the next big deal. Butte was mining steadily. The smelters were turning out copper by the carload. The Lewisohns and minor agencies were selling the product to a world in need of more and more copper for more and more electrical development. Public confidence had been restored.

Amalgamated was ready to extend its grasp upon Butte. The first scheme had worked so marvelously to the profit of its promoters that there seemed no reason why it should not be repeated. The first resources of the company included only the Anaconda properties, and the minor Parrott, Washoe and Colorado holdings. The Boston and Montana, the Butte and Boston, the Clark properties, and the interests of the smaller independents, including F. Augustus Heinze, were still outside the trust.

Rogers put the necessary pressure upon the Boston interests and drew them into the maw of Amalgamated, expanding its seventy-five million dollars capitalization to one hundred and fifty-five million dollars to cover the new issue of stock. That issue was floated in 1901 with a fanfare similar to that for the original issue. It was seized upon as eagerly as was the first issue. Again it was a simple matter to pay for the Boston properties out of the proceeds of that flotation, and without taking a cash dollar out of the pockets of the promoters. The public paid the bill, and when the market quotation of the stock soared to one hundred and thirty congratulated itself upon its apparent profits. But not for long. With the new profits safely consolidated by the promoters, the Amalgamated dividend was cut, and the market value of the stock plunged from a height of one hundred and thirty to a depth of thirty-three. That was less than half its valuation at its previous low, before the great rich mines of the Boston group had been added to its assets.

Thousands of investors were ruined. According to Lawson, press agent for the whole fantastic affair, basing his statement upon what he asserted was "personal knowledge," thirty men were driven to suicide, twenty were impelled in their extremity to crime which put them behind prison bars.

But Rogers, Rockefeller and Stillman remained in good repute. It was a far step in the forty years of advancing civilization from the crudities of a Montana mining community which hanged twenty-one men in thirty days for the theft of a few hundred thousand dollars and the murder of a few score men.

A quotation from Thorstein Veblen's *The Theory of the Leisure Class,* may be pat at this point. It is of timely interest that Veblen's sociological theories recently have been offered in simpler language as the original theories of some of the more vocal brain-trusters of the New Deal. Veblen, shortly after the Standard Oil group had made its second great killing in Amalgamated, wrote as follows:

"It is also a matter of common notoriety and by-word that in offenses which result in a large accession of property to the offender he does not ordinarily incur the extreme penalty or the extreme obloquy with which his offense would be visited on the ground of the naïve moral code alone. The thief or swindler who has gained great wealth by his delinquency has a better chance than the small thief of escaping the rigorous penalty of the law; and some good repute accrues to him from his increased wealth and from his spending the irregularly acquired possessions in a seemly manner. A well-bred expenditure of his booty especially appeals with great effect to persons of a cultivated sense of the proprieties, and goes far to mitigate the sense of moral turpitude with which his dereliction is viewed by them."

Veblen continued with similar expressions of similar and allied theories until his connection with the Rockefeller-endowed University of Chicago was severed. In the meantime the Standard Oil magnates who had promoted Amalgamated Copper were meeting with an opposition in Butte far more practical. The opposition of F. Augustus Heinze was, to be sure, based upon a theory, but no sociological theory. His was the theory of the "apex," and he was the man to make it practical, to his own ends.

Heinze had promised Colonel Bigelow of the Boston companies a battle. That battle was well under way, to the delight of Butte, when Heinze issued from his corner after one of the early rounds to find that his opponent had miraculously developed into a brutal bruiser with all the speed, footwork and science of an Abe Attel, with the weight and reach of a Primo Carnera, and the stamina of a Battling Nelson. Heinze had undertaken to beat a Boston company, and now found himself fighting the greatest, most powerful, most ruthless and conscienceless trust on earth.

Such a discovery would have sent most fighters scurrying through the ropes. Not F. Augustus Heinze. He waded jauntily into the new and colossal opponent, and boasted of his prowess. That Heinze was not knocked kicking at once was probably due to his comparative insignificance. The Standard Oil group could not conceive of any petty independent mine operator having the temerity to oppose their purposes. William A. Clark was a man who interested them, not for his fighting ability, but for the valuable mining and smelting properties he still owned in Butte. They would deal with Clark. But Heinze? Phooey! They would crack him like a louse under a horny thumbnail.

CHAPTER XVI

DAVID MEETS GOLIATH

HEINZE's early contests with the Boston companies had been fought upon the issue of whether the ore bodies in his Rarus Mine had their apex in the adjoining Michael Davitt Mine, Boston owned, or whether the apex was in Heinze's property and thus gave him title to the Michael Davitt ore. That battle reached the courts in March of 1898. That was before the Amalgamated Copper Company had been formed, but while rumors of its forthcoming organization were rife in copperland and in New York and Boston.

The Boston corporation sued Heinze and his Montana Ore Purchasing company in the federal court in Butte to establish their title to the Michael Davitt ore, which Heinze claimed under the apex law, and to Heinze's Rarus ore, which the Boston company claimed under the same law. The presiding judge directed a verdict for the Boston plaintiffs. Heinze's personal popularity and political and economic influence in Butte promptly revealed itself. The jury defiantly took a chance on going to jail for contempt of court by refusing to return the verdict dictated.

Before the case was set for trial again, in Helena, the line-up of personalities behind the suit was considerably changed. What Heinze described as "A Bill to Turn the Government of the State of Montana Over to the Standard Oil Company," had been passed by the same legislature which had been discredited as venal in its election of William A. Clark to the Senate.

The bill permitted a corporation, with approval of two-thirds of its stockholders "to sell, lease, mortgage, or exchange for the whole or part of the capital stock of any other corporation, whether domestic or foreign, . . . the mining ground, quartz mills, smelters, concentrators, reduction works, or other property or assets."

Governor Robert B. Smith promptly vetoed it. He described it as "An Act to Reverse the Supreme Court and to Encourage and Facilitate the Formation of Trusts and Combines and to Legalize the Confiscation of Private Property for Private Uses in the State of Montana." His veto message declared that Standard Oil and the Rothschilds had sponsored the bill with the purpose of forming a copper trust. "It has been rushed through with whip and spur, and every corporate lobbyist is at its back. The haste with which it has been railroaded through both houses ought to be a warning. They (the Standard Oil Company) control already the lamplights of America, and by controlling the production of copper they propose to control the electric lights of the world."

That did not disturb the Montana legislature in the least. They promptly passed the bill over the governor's veto by a vote of fifty-two to eighteen. The *Boston Transcript* announced: "The Montana legislature has passed over the governor's veto the bill permitting the transfer of corporation assets . . . which will allow the Boston and Montana Company to transfer its property to the New York Company of the same name. Mr. Lawson says that 'the passage of this bill marks the end of the fight between the Boston and Montana Copper Company and Heinze, and is the only thing wanting to complete the great copper consolidation which will now be announced in a few days.'"

Mr. Lawson was unduly optimistic. But his association with Rogers and William Rockefeller was widely known, and his announcement at least served to impress Montana with the fact that the Standard Oil magnates were behind the bill in the Montana legislature and would control the copper trust. Incidental was the fact that an effort was being made to transfer the Michael Davitt suit between the Boston company and Heinze to an eastern court. But Heinze had learned the usefulness of efficient corporation lawyers. He blocked that move.

When the Amalgamated was organized, a month after Lawson's optimistic announcement in the *Transcript*, the Boston properties were not included. As far as Heinze could see he was still fighting no one more powerful than his original antagonist, A. S. Bigelow, who had been head of the Boston companies for years.

When the Boston companies were conspicuously omitted from the amalgamation, Heinze promoted a suit designed to break up the affiliation of the Butte and Boston and the Boston and Montana companies, and to prevent subsequent transfer of the Boston and Montana to the Amalgamated. This transfer was already being rumored as the next projected move toward completion of the world-wide copper trust.

John MacGinniss, Heinze's right-hand man, with one hundred shares of Boston stock to give him the necessary legal standing before the court brought the action before Judge Clancy to block transfer of the property to the Amalgamated. A detail of the litigation was the appointment of a receiver for Boston and Montana, which was paying more than one million dollars a year in dividends. Judge Clancy appointed Thomas R. Hinds, one of Heinze's lieutenants. The company promptly barricaded the properties to keep the receiver and Heinze out. The next day Boston and Montana dropped twenty-eight points on the Boston board.

There are various conflicting rumors of what happened after that. One is to the effect that Rogers and Lawson, unaware of what had happened in Butte, called Heinze into hasty consultation, and offered to make a settlement with him, which he declined. Another and probably more accurate story is that Rogers, having failed to lure Bigelow into the amalgamation at an absurdly low price for the Boston properties, decided that Bigelow and the Boston copper millionaires must be disciplined.

In any event, with the bottom falling out of the Boston companies' market, there was no reason why the promoters of Amalgamated should worry. The farther that stock fell in value, the easier it would be to drag its rich assets into the Amalgamated at a price which would be doubled as soon as it came under the control of such well-known money-makers as Rogers, Rockefeller and Stillman.

Bigelow, perhaps being under the delusion that he was fighting only Heinze, countered with an action in the Supreme Court of Montana. The court held the transfer void and dismissed the receivership. Hinds, who had been in legal possession for five days, turned in a bill for four hundred thousand dollars for himself, fifty thousands dollars for attorneys' fees, and thirty-one thousand dollars for his two assistants. Judge Clancy seemed to think that eighty thousand dollars a day for Hinds' services was a trifle high. He cut the wage to only forty thousand dollars a day, and assessed a total cost of two hundred and eighty-one thousand dollars against the company. Things were getting a bit rough in Montana.

But they were no rougher than in Wall Street. Bigelow was in for trouble. His stock did not rebound with the dismissal of the receivership. He knew his properties were worth the higher market value on the basis of their resources and earning power. He had a long and enviable record in Boston as a copper magnate. He threw his personal fortune of ten million dollars into the market to protect the Boston copper stocks. The fortune was engulfed, and with it the Globe National Bank, in which he was a leading figure. In that situation he capitulated, not to Heinze but to Rogers. The Boston properties had not yet been taken into the Amalgamated, but the handwriting was upon the wall.

Under these conditions the second trial of the Michael Davitt suit was called in Helena. Heinze, now experienced through the

Clark campaigns as a soap-box orator, keen in his understanding of human nature, wise in advance of his time to the subtle value of propaganda appealing to prejudice and emotion rather than to intelligence, prepared a black and sordid history of Standard Oil. He induced friendly newspapers to print it. He announced to the miners, who had always been his friends, that Standard Oil methods were to be used through the Amalgamated company to cut their wages.

The trial of the suit continued for forty-five days, and the jury returned a verdict for Heinze. The Amalgamated promptly demanded a new trial on the grounds of prejudice induced by the attacks upon Standard Oil. The appeal was granted, and both sides injoined from mining the disputed veins pending final decision. For a moment there was apparent peace. But only for a moment.

Prior to this unstable armistice Heinze had discovered, surveyed and secured title to a tiny, unrecorded, triangular fraction of ground on Butte hill, as mentioned in an earlier chapter. This tiny claim, containing only three hundred and seventy-five square feet of ground, approximately one-twentieth of the average city building lot, he incorporated into a company under the impressive and perhaps ironical name of Copper Trust. Heinze had a sense of humor.

The legal groundwork having thus been laid, Heinze boldly announced that the great Anaconda, St. Lawrence and Neversweat Mines apexed in that property, no larger than a fairsized living room. The boy from Brooklyn was feeling his oats. His legal counsel, which he was developing into a staff that eventually numbered thirty-seven lawyers, applied to Judge Clancy for an injunction restraining the three big mines from operation, pending determination of the title.

That was a tactical error on Heinze's part—one of the few recorded against him during his operations in Butte. He was no longer fighting local capital or even the great Boston copper magnates. He was up against a real copper trust now, and his two by four so-called Copper Trust received a swift kick.

H. H. Rogers, William Rockefeller, William G. Rockefeller, J. P. Morgan and other directors of the Amalgamated may not have had any knowledge of the details of mine surveying, mining engineering and copper smelting, but they had a truly great knowledge of human nature and the tactics of their own trade. Also, they employed the best legal counsel obtainable. Their reply to Clancy's injunction was swift and effective.

They closed down the mines, without a moment's hesitation, and thereby threw three thousand men out of work. And every one of the three thousand was advised that they had been thus summarily cut off the payroll by the arbitrary and unjustified action of Judge Clancy in supporting Heinze's attempt to steal a hundred million dollars' worth of ore through a trumped-up technicality. Three thousand miners, marching down the hill with their empty dinner pails in their hands, not knowing where or how they would fill those pails again, were ripe for riot. Provocative agents among them whispered that Clancy was to blame. The mob spirit stirred.

Friends hastened to Judge Clancy with the warning that three thousand miners, abetted by their wives and children, were out to hang him. Clancy promptly vanished from his usual haunts, and passed word back to the powers that he would revoke the injunction as soon as the Amalgamated lawyers could prepare the papers. By midnight it was done. The next day the three thousand men went back to work.

Heinze had lost that round. But with a return of his usual shrewdness in dealing with the working masses of Butte, he dodged any personal criticism from the men by taking up their battle on another line. And incidentally he still held the Copper Trust claim as a threat over the head of the Amalgamated.

With the support of Clark, who was then actively engaged in the campaign for election of a legislature which would send him to the United States Senate without an accompanying memorial charging bribery, Heinze announced the adoption of the eight-hour day for all workers in his and Clark's mines. That went so well with the unions that Clark and Heinze were able to swing three-fourths of the votes of Butte, although they employed only one-tenth of the labor.

On Miners' Union Day in Butte, June 13, 1900, Clark and Heinze appeared together in a decorated carriage at the head of the parade. Amid the cheers of labor they demanded that the Amalgamated mines also adopt the eight-hour day. To further embarrass the enemy, while proving that their own hearts bled for the downtrodden workers, they demanded that the D. J. Hennessy Mercantile Company, "the company store," recognize the clerk's union and close the store at six o'clock instead of eight, thereby reducing the working hours there from twelve to ten.

The "company store" system had long been under bitter criticism in Butte as an instrument of peonage. Its practice there, as in most industrial centers dominated by a single company, had been to give credit to employees of the company, and collect direct from their pay envelopes at the end of each month. That system returned a large part of the payroll direct to the company, and at the same time kept the workers on that payroll under the thumb of the company.

Heinze attacked it in characteristic style. First he quoted the *Anaconda Standard,* which had gone over to the Amalgamated with the rest of the Daly properties, as announcing that the company would develop Montana to the high industrial efficiency and population of Pennsylvania. With that major premise he cited the squalor and misery of the Pennsylvania coal towns, charging that they had been brought to that condition by such methods as the Amalgamated and the Standard Oil companies used.

In that connection, to demonstrate the peonage made common by "company store" practices, he obtained and gave wide publicity to a sworn statement made by an employee of the Pancoast Coal Company in Pennsylvania:

"When I was six years and four months old I went to work in the breakers of the Pancoast Coal Company. I have worked nineteen years, every day I could get. I have never been on an excursion in my life. I have not drunk a drop of beer or liquor for five years, and for two years I have not smoked. I have practiced the closest economy in food, but I have never been able to accumulate $100 in my life. I have lived in the hamlet of F—— all my life. You and I know this has always been a company store town. We know in our hearts what that means, whatever the operators may say. Eleven years I worked for the Pancoast Coal Company and during those eleven years I swear here before the Omnipotent I never handled one cent of earnings in money."

It was quite a jump from Montana to Pennsylvania to get such campaign material, but Heinze had the agility to make it, and the nerve to use it, and expand it. That situation, he told the eager miners, could be expected from Standard Oil domination of Montana. That was what happened when company stores bled the workers of their wages. No widow of a miner killed in pursuit of his arduous and dangerous calling could start a tiny business to keep her children from starvation. No small shop-keeper could compete with the company store, not because the company store sold more cheaply—far from it—but because the company store could and did give credit to company employees for the full amount of their wages, and then seized the wages, leaving nothing to be spent elsewhere.

Heinze was hitting below the belt, but he appeared to be hitting in behalf of labor. The working men, most of whom were not analytical, had no qualms about that. For that matter, neither did

William A. Clark, who had plenty of education and plenty of analytical intelligence. What Clark wanted was a seat in the United States Senate. To win that seat he must control the incoming legislature. He recognized in Heinze a tremendous power to win votes through the appeal of a demagogue.

Although Clark had opposed the Daly organization in the election of 1898 on the grounds that he was opposed to fusion and was spending his tens of thousands of dollars in an effort to keep the Montana democracy "pure," his campaign of 1900 was made in support not only of the anti-Daly Democrats but anti-trust Republicans, eight-hour Republicans, and Populists. Heinze was the man to bring such strange political bedfellows together. Heinze was the man to make them like it. Little matters of consistency never cramped Heinze's style or cooled his ardor. He had his issue: Montana against Standard Oil; Heinze for labor against the labor-exploiting monopoly; Heinze for himself against any and all opposition.

As far as Heinze's personal interests were concerned, Clark's ambitions were incidental. But Heinze was so constituted that he could ally himself with any cause which might advance his own purpose. As far as Clark was concerned, Heinze's ambitions were incidental, but Clark was so constituted that he could adopt any ally who might help him attain his ambitions. So they came together and together fought the battle of 1900 against the Amalgamated Copper Company which had fallen heir to the Daly political organization.

Their first move was to secure control of the state Democratic convention. When the county convention in Butte revealed that the Amalgamated, employing nearly ten men to Clark's and Heinze's one, was still in control, the Clark and Heinze delegates walked out, and sent their own representatives to the state convention. Clark was strong outside of Silver Bow and Deer Lodge Counties.

He seated his delegation in the national convention in Kansas City, which nominated Bryan for the second time against McKinley. That was an incident, a gesture. What Clark wanted to do was to control the Montana legislature. What Heinze wanted to do was to elect his own judges, Clancy and Edward W. Harney, in Silver Bow County, and incidentally give the Amalgamated a black eye.

Clark put up ample money for the state campaign. Noted entertainers were imported from New York to draw crowds to the political rallies. Heinze took the stump as only Heinze could do, and flayed the Amalgamated Copper Company and the Standard Oil Company as the source of every evil known to the common people, past, present and future.

No ramifications of alleged corruption were overlooked. The issue outgrew the state of Montana almost overnight and extended to charges of perfidy and corruption in the federal government. The *Northwest Tribune,* published at Stevensville, Montana, October 19, 1900, raised the local issue to national prominence under a flaring headline:

"STANDARD OIL TIMBER THIEVES IN MONTANA

"More than a year ago John W. Griggs, McKinley's attorney general, ordered United States Attorney W. B. Rodgers, of Helena, to bring suit against the Bitter Root Development Company and the Anaconda Copper Company for stealing timber from public lands in Montana. . . . June 23, 1900, John G. Thompson, acting attorney general, wrote to Mr. Rodgers, saying:

"'The trespass involved the cutting of 23,525.125 feet of saw logs on unsurveyed lands lying within the limits of the Bitter Root Forest Reserve. As requested by the secretary, you are directed to institute and vigorously press to trial civil suit against said companies (the Bitter Root Development Company and the Anaconda Copper

Mining Company) to recover the value after manufacture of the logs. Report promptly your action under these instructions. . . .'

"This is the reason (for the delay): The Standard Oil trust and the Amalgamated Copper trust are the real corporations that control the lumber interests of Montana. The smaller timber-stealing corporations are the pigs of these two hogs. These hoggish corporations control Thomas H. Carter, United States Senator from Montana; and Tom Carter appointed W. B. Rodgers as United States attorney. Carter is the hired man of the Standard Oil and Amalgamated Copper trusts; he admitted that to F. Augustus Heinze last spring. W. B. Rodgers is Carter's hired man. And the Standard Oil and Amalgamated Copper trusts say the big timber-stealing corporations must not be sued. . . .

"The matter reached such a pass that H. H. Rogers, the new head of the Anaconda company, had to take the matter in hand. He summoned Senator Carter to New York, and Mark Hanna. H. H. Rogers is the political chief of the Standard Oil trust. The ramifications of this company are enormous. The banking trusts of New York are controlled by it. The great light and power companies of New York and New England are under Henry Rogers' thumb. Several of the great railroads are its vassals, and Rogers holds the political purse strings of the trusts that made McKinley president in 1896.

"H. H. Rogers told Mark Hanna that unless the suit against Anaconda company was abandoned, Mark Hanna would get no more money from the Standard Oil trust, the Amalgamated Copper trust, and their affiliated companies. He told him that in Montana he was spending millions to elect judges and a state ticket, and he could just as well carry the state for McKinley, but that if the Anaconda company were to be prosecuted for a million stolen government logs, McKinley and Mark Hanna should have no more money. . . .

"Mark Hanna went to Washington and laid all the facts before McKinley. . . . Thereupon the President undertook to straighten out the matter; it was fixed satisfactorily to Hanna, Rogers and Carter.

"So the people of Montana can see the tremendous power and influence of H. H. Rogers in national politics. And they can see how Carter is running the federal offices in Montana to persecute his foes and to shield his friends."

Along that line Heinze conducted the campaign, in a manner which held the interest of business and political leaders in New York, Boston and Washington as well as that of Butte. The Heinze battle had become an incident of national note.

Speaking before a crowd of thousands in Butte, Heinze declared that although there were six factions in the local field, there were only two parties—those who favored Standard Oil domination of Montana and those opposed. The candidates of the so-called Republican party and of the independent Democratic party, he said, alike represented Standard Oil. Every candidate on the Republican state ticket, he asserted, had been put there by Senator Thomas H. Carter, whom he branded as a hireling of the trust, and William H. DeWitt, attorney for Standard Oil. Every candidate on the independent Democratic ticket, he said, had been put there by John R. Toole, Standard Oil and Anaconda lobbyist.

"The Amalgamated Copper company, the Anaconda Copper company, the Boston and Montana company, the Butte and Boston company, and every copper company in Silver Bow county—outside those controlled by the Hon. W. A. Clark and myself—is absolutely dominated by Mr. William Rockefeller and H. H. Rogers, president and vice-president respectively of the Standard Oil company."

The mob howled approval. Heinze went on to state that he had met Rogers at the oil magnate's request in New York in 1899, and that Rogers said he was planning to combine the Butte mines, and later the mines of the world, with five hundred million dollars capital, to form a trust which would equal the Standard Oil trust, which had been able to increase its earnings fifteen million dollars a year on one occasion by the simple process of increasing kerosene prices arbitrarily by three cents a gallon.

"Mr. Rogers explained that situation to me, and I declined to have anything to do with his scheme."

Again the mob howled approval. Before a similar gathering at Missoula a week later Heinze elicited similar approval by an appeal to the popular admiration for Marcus Daly. Daly, said Heinze, probably expected to manage the Amalgamated as he had managed his own Anaconda, but soon found he was overruled by the executive committee of three. "After two or three months' residence here, when Mr. Daly had failed repeatedly to carry out his plans, he left here in August, 1899, a heart-broken, disappointed man . . . and has not since returned. In the interim his trusted lieutenants and men who have been with his corporations for years have slowly and gradually been replaced by the minions of the Standard Oil Company."

To reveal how the Standard Oil Company was feared by all who knew it, and perhaps to bring himself a little more prominently into the drama as the David who had defied this Goliath, Heinze asserted that he had heard himself pointed out with pity and astonishment in New York as "the man who is fighting the Standard Oil Company." More howls of mingled rage and admiration.

Heinze reveled in the battle. In a speech at Helena he proceeded to support his charges of Standard Oil efforts to dominate Montana by recital of the line-up of newspapers throughout the state. The *Montana Daily Record* of Helena, he asserted, was clearly spending far more than it earned legitimately in the support of Amalgamated candidates. In Anaconda, the *Standard,* pretending to be

Democratic, was in reality supporting Republican candidates pledged to Standard Oil practices. In Butte, he charged, the *Inter-Mountain's* sentiment and policy had changed with revealing abruptness to favor the Amalgamated when equipment was brought in and plans made for a rival Republican paper there. The rival, he said was then abandoned, and the *Inter-Mountain* supported the oil candidates. In Great Falls, the *Leader* "until recently having a hard struggle, suddenly indicates prosperity, and supports oil. The same could be said for the *Billings Gazette*, and the *Gallatin County Republican.*"

With that he began on personalities familiar to the voters of Montana whom he was addressing. He declared that United States Senator Thomas H. Carter had been and was accepting a retainer and salary from the Butte and Boston Company. He accused Senator Carter of acting as the trust's agent in the subsidization of the press. "Thomas H. Carter," he said, "did not put down the $500,000 that this campaign is costing, but the Standard Oil trust did." There was no equivocation about that. The voters to whom Heinze appealed loved it. Sensing his hold upon the popular fancy, Heinze continued with his own version of his labors in behalf of Butte and its workers, and his charges of attempted coercion by his enemies.

"Since the commencement of this lawsuit (the Michael Davitt case) the Hon. Lee Mantle, in the United States Senate, had a law introduced allowing the transfer of cases from Butte to Helena. I believe Mr. Mantle, who is a newspaperman and thoroughly acquainted with any legal matters, was induced to do this by the Hon. Thomas Henry Carter, and I think Mr. Carter had this done in his capacity as attorney for the Butte and Boston Company. The law was passed.. . . In the months of December and January of last year, after this property had been closed down for nearly three years, it finally came up in the United States Court of this community, and after a trial lasting some forty-five days before a judge and jury, the like of which has rarely come together, the jury found in my favor."

Continuing, he explained that charges of improper practices, court acceptance of affidavits instead of direct testimony and cross-examination, and so forth, had continued the case, and "I am still without a verdict."

The equally famous suit brought against him in the name of the Pennsylvania mine, one of the Amalgamated properties lying near his Rarus Mine, he said, had been brought by his enemies who sued for five hundred thousand dollars merely to tie up the Heinze property. When he retaliated with an action to quiet title, the plaintiffs procrastinated. Despite their opposition to a legal settlement, he said, he forced the issue to trial and, in January, 1899, won a verdict giving him the vein.

In the meantime, he said, he had started to buy adjoining properties to prevent more trouble on the apex theory, and had paid one hundred and fifty thousand dollars for control of the Nipper claim. There, he told the miners, he had uncovered rich ore where the original owners had found nothing to the glory or profit of Butte, themselves, or the workmen. He had carried the fight to the legislature and obtained a law through which he had managed to reopen the mine and put the miners back to work.

Only ten days later the Washoe Copper Company, also a unit in the Amalgamated, obtained another injunction against his operation in the Nipper shaft. He had foiled that effort to throw the men out of work. He had opened the mine at another spot, not covered by the injunction and kept the men on the payroll. Then the Parrott Company had sued, claiming that the ore of the Nipper apexed in the Little Mina, to the north. Every title, and every company, he declared was virtually the same, all being under Amalgamated control, yet each swore to a different set of alleged facts.

A dozen similar incidents, he asserted, were available to prove that the Amalgamated sought to rule or ruin Butte by ruling or ruining Heinze as the chief enemy of the trust and the leading friend of the people. He had paid two hundred thousand dollars for the Cora and Rock Island claims, and had found ore there where no ore had ever been found before. Yet W. J. Ladd, treasurer of the Boston and Montana Company, had bought the adjoining Silver Queen property and obtained an injunction against Heinze on the apex theory. Later the Butte and Boston closed him down on the charge that the Cora and Rock Island ores apexed in its Gray Rock.

All this Heinze cited in an effort to prove conspiracy by the Standard Oil interests, a conspiracy which he declared was stifling Butte's production and restricting the prosperity of Butte's workers. As a further appeal to the voters, he declared that his Montana Ore Purchasing Company was producing four million pounds of copper a month, and employing thirteen hundred men, but that if it were not for the conspiracy against him he would, within a year, be equaling the Anaconda's production and employing as many thousand men as the Anaconda employed. This conspiracy, he asserted, had blocked his efforts, to the extent of twenty-five injunction suits, to bring prosperity to Butte.

The peroration of his campaign came on November 6, 1900, when he stood on the steps of the court-house in Butte and addressed ten thousand men. "Rockefeller and Rogers have filched the oil wells of America, and in doing so they trampled on every law, human and divine. They were ruthless in crushing every obstacle in their path. They wrecked railroad trains and put the torch to refineries owned by their competitors. They entered into a conspiracy with railroads by which competitors were ruined and bankrupted. Sometimes they were caught in the act, but they bought the judges and saved themselves from prison stripes and punishment. And it is the same Rockefeller and the same Rogers that today seek to get control of the executive, the judiciary and the legislature of Montana.

"It is Rockefeller and Rogers, the Standard Oil barons, who are responsible for the prostitution of the press of this state. . . . The gold which Senator Carter and his henchmen are scattering over the fair state of Montana, to prostitute its electorate, is Standard Oil gold. . . .

"My fight against Standard Oil is your fight. . . . We stand or fall together. . . . If they crush me tomorrow they will crush you the day following. They will cut your wages and raise the tariff in the company store on every bite you eat and every rag you wear. They will force you to dwell in Standard Oil houses while you live, and you must be buried in Standard Oil coffins when you die. . . . Let them win tomorrow and they will inaugurate conditions in Montana that will blast its fairest prospect and make its very name hateful to those who love liberty. . . ."

"Napoleon! Napoleon!" the crowd shouted in frenzy.

CHAPTER XVII

CLARK LEAVES HEINZE TO HOLD THE BAG

BRYAN, Clark and Heinze swept the state. Clancy and Harney, Heinze's judicial candidates, were seated safely on the bench of Silver Bow County, where more than a score of suits against Heinze and more than a score of counter-suits against various units of the Amalgamated were pending. Clark's personal ambition of years was assured of attainment. He controlled a legislature of which a majority was promptly to send him, completely "vindicated," to the United States Senate. And thereupon he sold out to the Amalgamated, turning over to the trust a large part of his most coveted property in Butte.

Heinze was left to hold the bag. It was a bag as full of trouble as if stuffed with rattlesnakes. But there were many millions of dollars in the bottom. All that was necessary was to reach in, barehanded, take out the snakes one by one, crack their heads off, and get down to the money.

There was no doubt that there was money in the bag, but Heinze faced serious difficulties in getting at it. With the defection of Senator Clark he found himself virtually without financial support. His chief assets were imponderable. He owned the two judges of the Silver Bow courts, Clancy and Harney. He owned a few favorable court decisions in various suits to establish his title to ore beneath the Michael Davitt, Pennsylvania and other shafts owned by the Amalgamated. To be sure, several of these decisions were on appeal to higher courts. He owned three hundred and seventy-five square feet of his Copper Trust claim which he declared gave him apex rights to the richest properties of the Amalgamated company. He owned control of the Rarus and the Nipper. He owned a smelter capable of handling far more ore than he could take from those mines. Unless it ran near capacity this smelter must run at a loss. He owned a wild and woolly propaganda newspaper, the *Reveille,* which was a political asset but an economic liability. He enjoyed the favor of the labor unions, representing nine-tenths of the voting power of Butte. He had a dozen lawsuits and two dozen lawyers, and a payroll of thirteen hundred and fifty men to support. But he had little cash.

In this situation, at the age of thirty-one, the youth who had entered Butte only eleven years earlier as a mine surveyor at five dollars a day, found himself alone, committed to war against such men as Henry H. Rogers and William Rockefeller who were backed by inexhaustible millions grabbed through their organization of the Amalgamated Copper Company and more millions available from their Standard Oil resources. It seemed to be F. Augustus Heinze against the financial world. He did not hesitate.

The first thing he needed was ore—ore to turn into copper and so into money through his smelter. No scruples could stand in the way of that necessity, and Heinze had none. He had been enjoined from mining the Michael Davitt ore, pending decision as to title. He had made arrangements to purchase the Minnie Healey Mine. He had proved himself as quick-witted and quick-footed and unscrupulous as his gigantic adversaries.

In one of the early suits brought against him by the Amalgamated he had faced the necessity of putting up a seven-hundred-thousand-dollar bond. Amalgamated counsel knew his precarious financial condition after Clark's defection. They planned to win their action by the simple process of preventing Heinze from posting the bond. Heinze knew a trick as good as that. He walked into his offices in Butte one day and fired five employees. Secretly, he gave them twenty-five thousand dollars and secretly they went to Wilmington, Delaware, where they organized the "Wilmington Bonding and Casualty Company." The company issued just one bond—a bond for seven hundred thousand dollars to F. Augustus Heinze. When trickery was afoot it was clear that Heinze could hold his own.

The battle for the Minnie Healey Mine is a classic example, known to every mining man and most stock market men who have been active in the United States as far back as 1900. Miles Finlen, a friend and associate of Marcus Daly in the development of Butte, had obtained from Daly a lease on the Minnie Healey, lying a third of a mile northeast of Heinze's Rarus, on the southeastern slope of Butte hill. Finlen was to sink the shaft two hundred feet deeper in search of ore. The terms of the lease included an option to purchase for one hundred thousand dollars at the end of two years. Daly evidently had little confidence in the value of the Minnie Healey.

Finlen expended fifty-four thousand dollars in development work, but failed to find any ore worthy of the name. The mine appeared worthless, but not to Heinze. He had an idea that he could trace the ores of the Piccolo and Gambetta claims of the Boston and Montana into Minnie Healey ground. That would be worthwhile, whether or not there was any ore at depth in the Minnie Healey. Heinze offered to take over Finlen's option and lease for the fifty-four thousand dollars which Finlen had already sunk in the property, and to pay in two annual installments of twenty-seven thousand dollars each.

In the suit which soon developed, Heinze testified that Finlen jumped at the offer. Other witnesses asserted that Finlen went to the Butte hotel, bragged of his bargain, and bought champagne to celebrate. Judge John J. McHatton, Heinze's attorney, drew the papers and called Finlen to sign. Finlen, according to the story that came out in the course of the subsequent lawsuit, protested some feature of the agreement. According to Heinze witnesses, he said:

"You go ahead anyway; I'll turn the property over to you. When I come back from the East I'll sign the papers. My word is as good as my bond."

Finlen's subsequent testimony was that he had given Heinze temporary possession only to allow him to make a survey in support of his apex theory, with a promise that an option of purchase should be arranged later if a price were agreed upon.

On his eastern trip Finlen learned that plans had just been completed for taking the Boston and Montana and allied properties into the Amalgamated. Those properties did not include the Minnie Healey, and when Daly heard that Heinze was in possession of the latter property he gave Finlen peremptory orders:

"Go back to Butte and get Heinze out of the Minnie Healey. Why, if that young buccaneer gets hold of the Minnie Healey he'll have us all on the hip. He'll claim ownership of every pound of copper in the Piccolo and Gambetta."

But by the time Finlen returned to Butte, Heinze had struck a huge rich body of ore in the Minnie Healey, and had no intention of getting out. He had discovered ore where even the internationally famous ferret of ore, Marcus Daly, had failed to find it. His original intention of using the Minnie Healey as a basis of lawsuits to claim ownership of the rich Piccolo and Gambetta could be put aside, but not forgotten. He would rush the Minnie Healey ore through his smelter, keep his thirteen hundred and fifty men on the payroll, and provide himself with the ready cash greatly needed to finance his continuing war with the Amalgamated.

The Amalgamated had a different idea, which had been put before Finlen. When he returned to Butte he refused to sign the papers which had been prepared by Heinze's counsel. Instead he ordered Heinze to get out of the mine, which Heinze's development work had increased in value from an estimated one hundred thousand to millions of dollars.

Heinze refused. Finlen assembled a force of miners and advanced to seize the property. Heinze was acquainted with such tactics. He was not limited, morally, intellectually or physically to vicarious battles in the courts. He was a tough hombre who kept himself in perfect physical condition by road work as regular as that of a prizefighter. He could strike a blow with his fist as well as through his attorneys. He could lead his men in hand-to-hand encounter as well as direct them in the varied tactics of his varied warfare. He was advised that Finlen was on the way to seize the mine by force, and he met the attack with an armed guard which ran the Finlen cohorts off the property.

Then Finlen sued for possession, but soon sold his whole interest, including his rights in the suit, to the Amalgamated. When the case came to trial it was before Heinze's judge, Edward W. Harney. Harney had been a cowboy in Nebraska before he became a lawyer. He was of a type quite different from Judge Clancy. He lacked Clancy's stubborn indifference to the professional scorn of the nation's leading lawyers engaged in promotion of the Amalgamated interests but had equal loyalty to the man who had raised him to the bench. He was a far more able judge than Clancy, but he had one serious weakness. He was too fond of hard liquor. Still, that weakness had not injured his professional reputation up to the time the Minnie Healey suit came before him.

The case did not come to trial until after Marcus Daly's death and William A. Clark's elevation to the Senate and alliance with the Amalgamated. Heinze was alone against the copper trust when Harney gave him his first great victory—a decision favoring him on all points in the fight for the Minnie Healey.

But a little thing like an adverse court decision could not stop the Amalgamated. Their lawyers, already famous in the courts and corporations of New York, had learned new tricks in the bribery- and scandal-ridden state of Montana. They had upset Heinze's ore wagon before when they had set aside the Helena court's decision in the Michael Davitt case on the grounds of prejudice induced by Heinze's vicious attacks on Standard Oil. Now they purposed to upset it again. To do so they prepared affidavits asserting that Harney had accepted a bribe from Heinze to render a decision in his behalf.

That upset, not an ore wagon, but an apple cart laden with very wormy fruit. The scandal which resulted was heard from New York to San Francisco. Virtually all metropolitan newspapers throughout the country gave it space. Several magazines of national circulation subsequently printed its details as written by their investigators. The Heinze-Amalgamated battle was big enough to fill page after page of space in *McClure's, Current Literature, Donohoe's, Leslie's, Everybody's.* The fact that some of these magazines have since fallen from their high estate was not due to any action for libel or defamation of character induced by their reports of bribery, chicanery and fraud in the contest going on in Butte. Their stories stood, without effective contradiction. Written by trained and disinterested observers, after the heat of battle, they may be summarized as the best available record of those dirty days.

The dirt started to fly when Charlie Clark, eldest son of the Senator, the same son who managed his father's political campaigns, invited Jesse B. Roote, Judge Harney's personal attorney, to his home and warned him that Harney was in for trouble. Roote happened to be a law partner of William A. Clark, Jr., brother of Charlie Clark. That was a connection made before Senator Clark had gone over to the Amalgamated. Now the Clarks were on the Amalgamated side, against Heinze and all his works. But still Roote was Harney's counsel.

In substantiation of his warning, Charlie Clark told Roote that Arthur J. Shores, chief counsel for the Amalgamated in the action before Harney, had said they would ruin Harney unless he confessed a bribe from Heinze. To be sure, evidence could be, and was, adduced to indicate that the Amalgamated had offered a bribe, but that made no difference. Roote promptly set forth to find Harney and convey the warning. The judge was not on the bench. He was not in his chambers, or in his home, or in the barrooms.

But Amalgamated counsel knew a few things about Harney that even his personal counsel did not know. Leave it to the Amalgamated. As Henry H. Rogers had told Thomas Lawson, even before his flotation of the Amalgamated, "Standard Oil has its own methods of doing things."

So the Amalgamated had learned of an attachment between Judge Harney and Mrs. Ada H. Brackett, a public stenographer in Butte. The company was well aware that Harney had been elected largely through Heinze's support, and even while trying the Minnie Healey case before him was busily seeking evidence which it might use to charge him with corruption in the event of an adverse decision. The judge's association with the stenographer seemed to be the best point of attack. Captain D'Gay Stivers, of counsel for the Amalgamated, happened to have an intimate friend named George B. Dygert, an attorney, who was acting as go-between in clandestine correspondence between Mrs. Brackett and Judge Harney.

Whatever the validity of its subsequent charges—and they were never proved—it is evident that no sneaking, eavesdropping detective work, even to the extent of blackmail, was too low for the great trust to adopt in pursuit of its ends. Stivers induced Dygert to obtain, for potential use by the Amalgamated, copies of letters written by Mrs. Brackett to Judge Harney, and even some of Harney's replies.

Stivers knew where to look for Harney when Harney's own counsel, Roote, was unable to find him to give him Charlie Clark's warning—or threat. The Amalgamated attorney found him, near midnight, in the home of Mrs. Brackett, and bluffed him into going straightway to a meeting with the company's attorneys in the Thornton Hotel. That was "a night of spurs."

Stivers and Shores sat down to the task of coercing Harney. Roote was advised of the meeting by J. M. Kennedy, a lieutenant of Heinze, who had seen the men enter the hotel. He forced his way into the conference; ostensibly for the protection of his client. Charlie Clark was called to aid in the deal. Counsel for the Amalgamated argued, threatened and offered inducements to Harney to confess that he had accepted money from Heinze to influence his decision.

The bids for an affidavit from Harney stating that he had accepted sixty thousand dollars from Heinze advanced steadily from fifty thousand dollars to one hundred thousand dollars. In addition, Harney was assured that the affidavits against him prepared by Amalgamated counsel would be destroyed without filing. Charlie Clark acted as intermediary, passing back and forth between the room occupied by Shores and Stivers and that in which Harney was undergoing his ordeal.

But Harney persisted in his denial that he had been given a cent, or otherwise improperly influenced by Heinze. Clark raised the bid to two hundred and twenty-five thousand dollars, "being empowered to do so by the lawyers," as he asserted in an affidavit later.

Still Harney stood pat. If he was bought, he was going to stay bought. Shores, Stivers and Clark decided to call in Mrs. Brackett, in the belief that twenty-five thousand dollars might win her influence to bring Harney around. Kennedy, at the keyhole, heard this plan and hurried away to Mrs. Brackett's home to advise her. The lady had been a beneficiary of Heinze's political influence to the extent of an appointment as stenographer to the legislature, and was believed to know upon which side her bread was buttered. Kennedy's opportunity seemed clear, his task simple.

Mrs. Brackett admitted him, although the hour was two or three in the morning. Before he could reveal his whole plan, however, there was another ring at the bell. Kennedy stepped into hiding and Stivers was admitted. Kennedy later asserted that he heard Stivers make a proposition to Mrs. Brackett to take care of her and Harney for life if she would influence Harney to do as the Amalgamated desired.

Mrs. Brackett seemed to listen to reason. When Stivers asked her to go with him to the hotel where Harney was being put under pressure, she asked him to go get a woman friend to accompany her. The lady had a reputation to preserve. When Stivers had gone on the errand Kennedy stepped out and urged her not to go, but to insist that they come to her with Harney, and their proposition. In the meantime, he explained, he would install a stenographer and witnesses in the adjoining room to take the evidence. Before the plant could be arranged, however, Stivers returned and Mrs. Brackett accompanied him to the hotel. There she was allowed to speak privately with the sweating judge. If she had any influence it was not apparent. Harney continued to deny that Heinze had bribed him.

Wearying of the futile ordeal Charlie Clark took things characteristically in his own hands "Hell," he said, "I'm not going to stay here all night. Call it two hundred and fifty thousand dollars, Harney."

Harney remained defiant. The meeting broke up. Ten million dollars or more slipped out of the grasp of the Amalgamated. Four hours later the affidavits were filed charging Harney with accepting a bribe from F. Augustus Heinze.

Impeachment proceedings were brought against Judge Harney by Amalgamated agents who pushed the following resolution through the legislature: "Resolved, That Edward Harney, a judge of the second judicial district, be impeached of high crimes, misdemeanors and malfeasance, and that five managers be elected by this house of representatives to prepare articles of impeachment, present them at the bar of the senate, and prosecute the same."

The five managers thereupon were selected and went to work, examining witnesses, obtaining affidavits and so forth. Again bribery charges were in the air. It was a period when very little could be done in Montana politics or business without dirty money being one of the chief subjects of conversation in the clubs, barrooms, lobbies and streets. Harney, ably supported by Heinze's powerful legal staff, tried to prove conspiracy by the Amalgamated.

Mrs. Harney, sticking to her husband through thick and thin, despite the documentary evidence which was produced to indicate his susceptibility to Mrs. Brackett's charms and influence in favor of the Heinze group, offered a strange story under oath. According to this story, on one occasion she started for a walk in the country with her husband, and before they got out of town they were met by a man known by both to have affiliations with the Amalgamated. Two or three others joined the party and walked ahead with the judge while Mrs. Harney followed with the Amalgamated agent.

Passing a beautiful residence, the agent remarked, "What a beautiful house that is."

"It is beautiful," said Mrs. Harney. "How I should like to have one like it."

"Would you?" said the agent. "You can if you want it."

"What do you mean?"

"Why, a woman can generally have what she wants, and I am sure you can. Besides, if you and the judge did not care to build in this country, you might choose England or France."

Indignantly, Mrs. Harney said, she rejected the suggestion of bribery, and had done nothing more about it until her husband had been accused, and she decided the incident should be reported to indicate that the Amalgamated had attempted vainly to do the same thing which they now accused Heinze of doing. Of course the story could not be substantiated, and nothing came of the incident, but it was in the record.

Before the impeachment proceedings were started, Judge Harney overruled a motion for a new trial and went east on his vacation. Upon his return, attorneys for the Amalgamated presented a bill of exceptions for him to sign. It was couched in such language that the judge committed the two attorneys to jail for contempt of court. Charlie Clark thereupon went to New York for an unlimited vacation, and D'Gay Stivers sought much-needed rest and change in Alaska.

The outlook for Harney in the impeachment proceedings improved. Heinze still had considerable influence in the legislature, despite Senator Clark's defection. The impeachment committee begged to report that it found evidence of conduct unbecoming official dignity on the part of Judge Harney, but nothing sufficient to justify impeachment. It recommended that the resolution calling for impeachment and prosecution be tabled. That was done. Harney was cleared. Heinze and his lawyers were outstanding in the informal reception which followed immediately in the lobby.

Still, few persons either in Butte or Wall Street were certain that bribes had not figured extensively in the Minnie Healey suit. The Amalgamated appealed the case to the Supreme Court of Montana. That court sent the case back to Judge Clancy for retrial on the grounds of improper conduct on the part of the original trial judge, Harney. Clancy, as was expected, upheld Heinze's contentions on every point, finding that the verbal agreement of Finlen to sell the mine to Heinze for one hundred thousand dollars was valid, and that legal title was therefore with Heinze. Again the Amalgamated appealed. And this time the Supreme Court upheld the decision of the lower court.

It began to appear that Judge Harney's decision was sound, whether or not he had ever been given a bribe by Heinze, or offered a counter-bribe by the Amalgamated. The case which had been in the courts for nearly six years was settled, and Heinze had won against all the forces and the millions of the Standard Oil magnates' copper trust.

Prior to that, however, an action was filed for disbarment of Attorneys Shores and Stivers, of the Amalgamated.

The break between Heinze and the Clarks was wide open. Their enmity, since William A. Clark's defection to the Amalgamated, had become as bitter, and even less dignified, as the twenty years' war between Daly and Clark which had preceded it. Heinze lost no opportunity to take a wallop at Senator Clark, either above or below the belt. His scurrilous little propaganda newspaper, the *Reveille,* was a brass-knuckled fist for this purpose.

Probably never in the history of American journalism has there been a more violent and unrestrained publication. Though no journalist, Heinze revealed to the newspaper world what could be done when a man with a printing press and funds really desired to exert himself in the way of cartoons, headlines and unbridled invective. His cartoons covered half the front page, and stopped at nothing. The *Reveille* was the blazing example of the point to which civilization had advanced in Montana in less than forty years from the establishment of the territory's first newspaper, the *Madisonian,* in Alder Gulch.

In the political campaign which Heinze was waging against Standard Oil simultaneously with his court campaigns, one outstanding half-page cartoon had depicted John D. Rockefeller, in turban and draperies, cross-legged on the cushions of a divan, with the lecherous grin of an Oriental potentate, looking with rapacious eyes upon Montana in the guise of a beautiful woman being presented by Senator Thomas H. Carter as an addition to the Standard Oil harem, while H. H. Rogers and others smirked in the background. That was one of hundreds, illustrative not only of the time and the campaign but of the *Reveille's* methods.

When Clark went over to the Amalgamated and left Heinze to hold the bag, half the young engineer's bitterness turned against his former supporter. Journalism in Montana was something to be entertained by, if not to respect. Witness this, from the *Reveille* of September 27, 1902, under a spreading headline:

W. A. CLARK INSULTS PRES. ROOSEVELT

"From the Butte *Miner* (Clark's newspaper) of Sept. 23.

" 'The recent incident in which the president shot a pig has revived a discussion relative to the disposition and tastes of the chief executive.

" 'The fact that Mr. Roosevelt could not restrain his appetite for hunting when President McKinley lay at the point of death may be cited as an instance of the vein of brutality running through the disposition of the rough rider.

"'His hunting experiences in Colorado were attended by exhibitions of brutality that would inspire contempt had they been attributed to a man of lesser note. . . . It is quite natural for the President to posture and pose before the public in situations calculated to impress upon the people the irrepressible energy of his make-up. He loses no opportunity to advertise his love of prowess and his belief in the principle that might makes right through his connection with the military arm of the government and his disposition to pose as a military leader and his efforts to excite applause through his experiences as the mighty Nimrod of the century.

"'These characteristics of our chief executive have caused a healthy and wholesome disgust in the minds of a very large majority of cultivated persons.

"'The man who can find enjoyment in the taking of life—mountain lions or possibly a pig when necessary—is not fit to represent the higher sentiment of a progressive people. Had Mr. Roosevelt been thrown in other currents of life he might easily have gravitated into the pig sticking department of a Chicago pork packery, rather than into the exalted position of the master of the Whitehouse.

"'The example set before the world by the chief executive in his capacity as a hunter will ever remain a source of disappointment to his friends. . . . It may be no fault of Mr. Roosevelt that his tastes run along the lines of wanton slaughter of animal life, but it is the misfortune of the American public to be represented in the Whitehouse by a man of that type.

"'Mr. Roosevelt's disposition is in marked contrast with the exalted Christian idea and humane and kindly purposes of his distinguished predecessor—the lamented McKinley.'"

With two such newspapers in Butte—Clark's, which would originate such an editorial outburst, and Heinze's, which would broadcast it to the world in an attack upon the other—anything might be expected. The *Reveille* lost no time in obtaining and printing Charlie Clark's deposition submitted in the disbarment proceedings brought against Shores and Stivers as an aftermath of the Amalgamated's action against Judge Harney. Headlines were glaring enough to catch the attention of everyone in Montana who could read:

ATTEMPTS TO CORRUPT THE JUDICIARY
CONFESSED BY CHAS. W. CLARK

STARTLING ADMISSIONS OF GUILT MADE IN THE SHORES-STIVERS DISBARMENT PROCEEDINGS
CALLOUS RECITAL OF CONSPIRACY ON THE PART OF CLARK AND AMALGAMATED COPPER COMPANY TO WRECK A HOME AND DESTROY A FAMILY—HARNEY REFUSED A QUARTER OF A MILLION

"Charlie Clark has confessed that he offered Judge Harney $250,000 for a lying statement concerning his decision in the Minnie Healey case.

"In two written depositions signed by himself, attested to by a notary public of San Francisco, and now on file in the office of the clerk of the court, Sam Roberts, Charlie gives the world one of the most remarkable stories of bribery that has ever been told.

"Besides confessing that he offered to the judge a quarter of a million dollars, he states that he heard Stivers offer Mrs. Brackett $20,000 and that he (Clark) made an additional offer of $5,000 to this woman if she would aid them in their designs.

"He confesses that he had conferences with A. J. Shores, John Forbis and other Amalgamated attorneys concerning the bribery of Judge Harney, and the Amalgamated people stated that they were willing to pay handsomely for anything they could accomplish with the judge.

"He confesses that when he spoke to Forbis and Shores he told them that he did not think Harney was a cheap man, and he believed $200,000 would be a reasonable sum to offer him.

"He confesses that Shores and Stivers arranged the famous meeting at the Thornton Hotel the night of the 5th of August, 1903, and that when he talked with Judge Harney he raised the stipulated amount from $200,000 to $250,000.

"He confesses that when he announced this to Shores the latter demurred to the amount, saying it was too much.

"He confesses that after remonstrating with Shores as to the amount the latter gave his consent.

"In general Charlie corroborates the testimony of Judge Harney and Attorney Roote as given before Judge McClernan and the judiciary committee of the lower house. . . .

"Last night from Maine to California every city in the United States was printing this story of the largest individual bribe ever offered."

Despite such evidence, Shores was cleared, and the proceedings against Stivers were dropped. Heinze, in fighting singlehanded the intrenched powers of the Standard Oil interests, was finding altogether too many things to do.

CHAPTER XVIII

WAR WITHIN THE EARTH

HEINZE and the Amalgamated were providing exciting entertainment for Butte, for Wall Street, for Boston and for the copper-consuming world. In effect, the opposing forces were keeping scores of keen-edged knives flying between them as a Japanese juggling team might do. Millions were the prize, death or dismemberment the penalty. In fact, one hundred and thirty-three lawsuits were pending in the courts at one time.

The Silver Bow county, state and federal courts in Montana were so crowded with the mining litigation that ordinary private suits were delayed as long as seven years. Heinze maintained a legal force which mounted at one time to thirty-seven attorneys, with his brother Arthur Heinze brought from a lucrative private practice in New York to take charge.

A few years later Heinze was to tell John Gillie, chief engineer for the company, that he had a band of wolves around him who would not let him quit. But that was after the smoke of battle had cleared. The fact was that Heinze did not want to quit. He wanted to win, and he was willing to go to any lengths to win. The opposing force was no more scrupulous than he. Comment nationally circulated through *Leslie's Magazine* at the height of the warfare sums up the situation with an editorial fairness and accuracy now emphasized by a third of a century in retrospect:

"It is hoped Mr. Heinze can be curbed by law without being crushed, even though he represents a type of business man both predatory and parasitic. For he has made a more effective resistance to the deadening influence of the Standard Oil domination than any other kind of man could possibly have done. His course has been without moral justification but not wholly without excuse. Indeed the most depressing feature of the conflict to an unprejudiced onlooker is the fact that if he had been a more scrupulous man he would long since have been wiped off the slate. . . . He has been willing to play the game unfairly and win on a foul."

Probably, in greater justice to Heinze, it should be said that the contest had taken on the character of a battle-royal in which a dozen hungry, jobless men are put into a ring at a cheap athletic club with the understanding that the one who survives will be rewarded with the job of janitor of the club. There were no Marquis of Queensbury rules. There were no more rules than in any rough-and-tumble mining camp fight. Without rules there could be no fouls.

The Amalgamated was represented by men, underground and in the courts, no more scrupulous than Heinze. They fought with one idea and one purpose only—to win. The end would justify the means. Only their chief of counsel in Montana, William Scallon, was handicapped in such a knock-down-and-drag-out fight by his personal honesty and respect for the ethics of his profession. And Scallon was to part company with the trust, sacrificing a fortune, before the warfare ended. Long before that time the blows and kicks were flying so fast and furiously that even a man of Scallon's ability and devotion to duty could not keep track of them.

So extensive and involved were the court actions that a special set of books had to be kept in the Amalgamated offices to keep them straight. While a dozen suits were pending or being tried, another dozen were being reversed or appealed. While a federal court was sending back one action to a state court for retrial, injunctions were closing down half a dozen mines. While a mine was closed, one side would discover that the other side was mining in property it claimed, in spite of the injunction. Court orders were being sought and granted to stop such thefts. Force was being used to prevent the execution of the orders. Ramifications were extending into the stock market and the seats of the mighty in New York.

Amalgamated shares which had been floated at one hundred, and climbed to a quotation of one hundred and forty even before they were issued, slumped to seventy-five. At that figure H. H. Rogers and his Standard Oil associates bought them in and boosted them to one hundred. With the flotation of the second section of Amalgamated, including the rich Boston and Montana and the Butte and Boston, the price climbed to one hundred and fifty. In the succeeding riotous trouble with Heinze it crashed to thirty-three. The financial world was feeling the battle of Butte.

Having won a decision giving him the Michael Davitt ore, only to have it blocked by the Amalgamated appeal and ensuing injunction, Heinze squirmed out of that impasse by organizing the Johnstown Mining Company and assigning his rights in the Michael Davitt to it. Asserting that the injunction applied only to Heinze's original company, the Johnstown company proceeded to block off with concrete bulkheads every opening from the Amalgamated's adjoining Pennsylvania mine into the Davitt. That was not mining. There was no injunction against that. But when it had been done, the Johnstown Company proceeded to run crooked crosscuts from Heinze's Rarus into the rich ore bodies of the Michael Davitt. One vein in the Davitt, called the Enargite, ran forty percent copper. Another, the Windlass, was almost as rich.

The Johnstown Company proceeded to loot them, hoisting the ore through Heinze's Rarus shaft and sending it to Heinze's smelter. To the man on the surface the ore appeared to be coming from Heinze's Rarus. As fast as the stopes in the Davitt were emptied of their rich ore they were filled with waste and cemented up. No one could prove that there had been ore there.

But workers in the Amalgamated's Pennsylvania could hear the blasts in what they knew to be Michael Davitt ground. The company managed secretly to get one of its own men into the Rarus to investigate. Following along one of the circuitous crosscuts the scout fell down an inclined chute, twenty feet deep, to where a group of miners was working under a Heinze foreman. Picking himself up, he began nonchalantly to brush the dirt from his clothes.

"Hello," said the foreman; "where did you come from?"

"From County Galway, sir," said the scout.

That won a laugh. The situation was saved, as far as the scout was concerned. He was allowed to depart with the conviction, but no proof, that Heinze men were mining in the forbidden Michael Davitt ground. When the incident was reported to Heinze he doubled the force which was stripping the Davitt veins of their wealth. Amalgamated lawyers by various and devious methods obtained enough information to warrant a court order that their engineers should be admitted into the mine for examination. But immediately after granting the order the federal court adjourned for an indefinite period.

When the Amalgamated investigators appeared at the shaft to enter the mine, Heinze's employees refused to allow them to go down. The court which had granted the order having adjourned, nothing could be done about it. Heinze continued to rush the ore to his smelter. The Amalgamated was desperate. No one in the history of American big business had ever so persistently defied and fought Standard Oil interests with their own methods.

But these interests were no more to be stopped than was Heinze. The Amalgamated ran a crosscut from its neighboring Pennsylvania mine into a Heinze winze which led up to the stopes in the Davitt. By that time the confusion concerning apex rights and geological formations and so forth had reached such a point that only the trained and experienced expert could follow the devious veins and workings. Even the experts could, and did, disagree. The underground workings were a labyrinth, hundreds of miles in extent, difficult for a layman who has never been in a great mine even to imagine.

A far-fetched illustration might help. Imagine the Empire State Building, the world's largest office building, completely buried in the earth. Assume that half of its rooms and corridors, some large suites, some whole floors, are filled solidly. Many of its corridors, instead of running straight, are curved and twisted, turning abruptly at odd angles or making wide arcs. Here and there at unexpected points are open shafts with ladders dropping from the corridors to levels perhaps a hundred feet lower. Here and there a great hole in the ceiling leads up to higher levels, accessible only by ladders. Narrow railroad tracks lead along some corridors. Mud and rock pave others. Plank chutes come down at many points to carry material, tumbling and roaring into bins for transfer to dump cars and hoisting to the surface. No one knows what lies below. No one is certain what is contained in the vast areas filled with solid rock. That is rough, limited, but for the layman a serviceable picture of the underground workings of the Michael Davitt Mine.

Multiply it by one hundred to include the scores of mines under a square mile of surface of Butte hill and you may have some idea of the physical environment in which the underground battles between Heinze's miners and the Amalgamated miners were fought.

So involved was the labyrinth that the Amalgamated had put skilled mining surveyors, operating with famous geologists, to work to define its holdings, its veins and its titles. Among them was one young engineer named Samuel Barker. Barker spent five years underground in the making of surveys and the preparation of data for the company's use in its many suits. He is still a practicing engineer in Butte, and is one of the sources of information for this statement of the nature of the hand-to-hand encounters which went on within the earth.

When the Amalgamated engineers were thwarted by Heinze's defiance of court orders to allow them to inspect the Rarus connection with the Davitt, they dug through the earth from the Pennsylvania. Their purpose was twofold: to stop Heinze's looting of Michael Davitt ore which he, as they, had been enjoined from mining and to obtain evidence of how much ore he had already stolen.

Heinze had no intention of allowing them to accomplish either purpose. When they broke through from the Pennsylvania into the Davitt diggings Heinze's men poured slacked lime into the pipes which blew fresh air into the workings, and smothered the invaders. They retreated for a time, and the Heinze men rushed their ore out to the hoist, and filled the looted veins and stopes with debris. The Pennsylvania miners returned with steam pipes and high-pressure water hose to drive out the enemy.

Their entrance had been blocked with debris. As fast as the Pennsylvania workers pierced the debris, avalanches of rock drove them back. Some were severely injured. Still they came on, attempting to clear the way with their steam and water nozzles directed against the suffocating clouds of slacked lime from the Heinze blowers. Some of the Pennsylvania men found another tortuous passage into the Davitt diggings and attempted to take the Heinze warriors in the rear. A Heinze rear guard discovered the approach from a lower level, and dropped a whole bucket of lime upon the head of the bold invader, knocking him cold. But as the Heinze man was about to throw another bucket he was knocked over by a high-pressure stream thrown by the Pennsylvania cohorts.

The Amalgamated top-sergeants led a new assault up the raise, attempting to shield themselves from hurled rocks and lime with wooden barriers lifted above their heads. The natural flow of air through this section of the mine was against them. Heinze men burned rags to sweep the invaders back with a stinking smoke which was a forerunner of gas warfare. The battle went on for days. Heinze men were as thick as rats in the diggings. The stopes were being looted more rapidly than ever, despite the battle. Pennsylvania men would find a bin full of rich ore when they made an advance, and by the time they retreated to that point under renewed pressure the ore would have vanished, snatched away through some secret passage.

The battle raged through a mile and a half of underground workings, according to a statement of one of the Heinze engineers later. "More powder was burned than in the Russo-Japanese War," the current war of that day.

"Hooey!" says Sam Barker. "Most of the powder burned was in demolishing looted stopes and veins to destroy evidence of the extent of the looting. Dynamite sticks with fuses long enough to allow plenty of time for escape by both sides were thrown, and occasionally exploded, but that was mostly horseplay. One of the favorite tricks by the Heinze forces to block the advance was the electrification of 'turn sheets.' 'Turn sheets' are the heavy metal plates sometimes placed where a drift or crosscut turns an angle so acute that the ore cars cannot make it on the tracks. They can run off the tracks onto these sheets, make the turn, and run onto the tracks again.

"The Heinze men ran electric wires to these 'turn sheets,' and charged them strongly enough so that any man who stepped on them would receive a shock, sometimes sufficient to knock him down, always sufficient to discourage him from further attempts to advance. Even so, it was bloody warfare, but not so bloody as the writers of thirty years ago sought to indicate."

The only two deaths clearly attributable to the warfare were those of two miners, Oleson and Dwel, killed by a blast attributed by the Amalgamated to the Rarus forces and by Heinze's *Reveille* to the Amalgamated. Incidentally, one of the widows eventually won a verdict for twenty-five thousand dollars damages from Heinze.

In the meantime the Amalgamated, getting nowhere in its underground warfare, appealed to Judge Hiram Knowles in a federal court to give the company an even break with the buccaneering Heinze. "There are eight bulkheads," said John F. Forbis, representing the Amalgamated in this action, "and if we are compelled to force our way under ground, it will be necessary for us to run a blockade. . . . If we cannot stop this, the Government might better abandon its courts and leave litigants to determine their rights by the shotgun. Your Honor should dissolve the order enjoining us from working this ground, and let us go in there and fight for our ore."

Judge Knowles granted the order. Heinze evaded service for days while his crew rushed the richest of the remaining available ore to his smelter, and wrecked with dynamite the stopes and veins from whence it came. The case was carried to the United States Circuit Court of Appeals in San Francisco. That court instructed Judge James H. Beatty of Idaho to go to Montana and settle the row. He found Heinze guilty of contempt in having ignored the original court order, and fined him two thousand dollars, and two of his superintendents five hundred each. More important, he ordered Heinze to admit the Amalgamated engineers for inspection. This time it worked. R. H. Sales, a leading expert and mining engineer for the Amalgamated, made the inspection. Testifying later, he said:

"When I went down, pursuant to the order of the court, to make an examination of the underground workings in the Rarus and Michael Davitt, I found that the Heinze miners had scooped out a horseshoe-shaped passage and had crosscut around until they had tapped the Pennsylvania's rich Enargite vein in property of the Butte and Boston. They left the face of the vein at the end of the Pennsylvania tunnel intact, so that upon a casual examination no one would know that the vein had been tapped twenty feet or so farther on.

"They filled in and concealed this horseshoe burrowing with boards and debris. How much ore did they get out of the Butte and Boston vein from that burrowing? If you mean how much altogether, I cannot tell you, because they drove me back with dynamite bombs before I had a chance to complete my examination and estimate. But I can tell you how much ore was extracted from that portion of their surreptitious cutting that I actually saw. It amounted to 7,500 tons. Its value would be somewhere between $500,000 and $600,000."

Heinze's *Reveille* told a somewhat different story under a banner headline:

U. S. COURT OF SAN FRANCISCO STAYS JUDGE BEATTY'S ORDER

AMALGAMATED INSPECTORS ORDERED TO LEAVE RARUS MINE IMMEDIATELY—HIGHER TRIBUNAL TURNS TABLES UNEXPECTEDLY ON STANDARD OIL COPPER TRUST—STORIES OF STOLEN ORE NOW PROVED MALICIOUS FALSEHOODS

The dispatch under a San Francisco date line asserted that the United States Circuit Court of Appeals had granted a writ of supersedas in the matter of contempt of F. A. Heinze and others in litigation between the Butte and Boston and Heinze's Montana Ore Purchasing Company. In that case Judge Beatty of Idaho, acting for Judge Knowles of Montana, said the dispatch, fined Heinze two thousand dollars, Josiah H. Terise five hundred dollars and Alfred Frank five hundred dollars with a proviso for remittment if Heinze granted the Butte and Boston the right to survey and inspect.

The *Reveille* made that announcement as a victory for Heinze. A week later it burst forth again, reciting what it asserted was a greater victory in the federal court in San Francisco:

HEINZE WINS AGAIN IN U. S. COURT
STANDARD OIL TRUST TURNED DOWN

This story cited an oral opinion of Judge Morrow of the Circuit Court of Appeals denying a motion to set aside the writ of supersedas. The motion had been made by Attorney Thornton for the Amalgamated, charging that Heinze was still violating the injunction, and quoting Attorney Forbis of the Amalgamated to the effect that Heinze had stolen three hundred thousand dollars' worth of ore. In that situation the court had stayed the order of Judge Beatty and instructed the Amalgamated inspectors to get out.

"Immediately," said the *Reveille* account, "the Amalgamated Copper Company resumed the blasting and dynamiting of the stopes, and the very evening after the order of the court arrived they exploded an immense quantity of powder by which two of their own men lost their lives."

That was Heinze's story as opposed to the Amalgamated's tale of a theft by him of more than five hundred thousand dollars' worth of ore. Whether or not it was accurate, Heinze had managed to confuse the issue, just as he had confused it when the Amalgamated charged him with having bribed Judge Harney, and he answered with sworn testimony indicating that the Amalgamated had offered an immense bribe and had been repulsed. The general opinion in Butte, however, was that Heinze had obtained upward of a million dollars' worth of ore from the disputed ground of the Michael Davitt.

Not so bad for Heinze, struggling desperately to keep his mines, mills, lawsuits, newspapers, judges and politics financed in his fight against the directors of the world's greatest trust. He could afford to pay three thousand dollars in fines for such a profit. It was worthwhile continuing, and he continued.

Within a year Judge Beatty was again called in to settle a new angle of the case. He heard testimony that Heinze had taken more than one million dollars from the Michael Davitt. This time he fined the lone opponent of Standard Oil twenty thousand dollars. There still remained a profit of some nine hundred and eighty thousand dollars for Heinze. He had not been on trial for theft. The charge was contempt of court. He should worry about a little thing like that.

Butte could laugh most rudely when eventually the Supreme Court decided that Heinze had never had a legitimate title to the Michael Davitt ore. He seemed to have obtained a million dollars' worth of it, and to have thrown that sum into his fight against the legitimate owners of the million.

And that was not all. Far from it.

CHAPTER XIX

ONE AGAINST THOUSANDS

THE fight for the Minnie Healey had been going along at the same time. When Judge Harney had decided that Heinze's verbal contract with Miles Finlen for lease and option on the Minnie Healey was valid, despite Finlen's denial, Heinze had gone energetically to work to trace the Minnie Healey veins into the workings of the Leonard and other Amalgamated properties. The immediate purpose was to tie up the Amalgamated through a suit claiming apex rights to their ores. But better than that, Heinze encountered in the Minnie Healey one of the richest ore bodies under Butte hill.

While the Amalgamated was bringing bribery charges in vain effort to set aside Harney's decision, Heinze was mining furiously in the Minnie Healey, turning ore from the disputed ground into money with which to fight the disputant.

Heinze was taking too much money from Minnie Healey and Michael Davitt ores. The copper trust could not tolerate this continuation of supply of munitions from its own stock to its privateering enemy. So, when it had appealed the Harney decision, and the State Supreme Court had sent the case back to Judge Clancy for a new trial, the Boston and Montana unit of the Amalgamated began tactics against the Minnie Healey similar to the tactics adopted by Heinze in the Michael Davitt.

The underground warfare was on a par with that in the Davitt. Heinze engineers tapped the water column of the Minnie Healey, coupled with the city water works, and turned the flood into Boston and Montana workings. Boston and Montana engineers promptly erected a barrier and turned the flood back through a crosscut eight hundred feet into the Minnie Healey shaft.

Heinze workers poured out of the lower workings of the Minnie Healey like rats out of a flooded cellar. Word spread through the town that scores of men had been trapped and drowned by the Amalgamated's quick turn of the Heinze attack.

Hysterical wives and children of the miners ran screaming through the streets to the Minnie Healey shaft. Fifteen hundred persons formed a wildly excited mob. Threats of lynching of Amalgamated officials arose. The mob turned toward the office of Wallace Corbett, an Amalgamated engineer who was reported to be responsible for the flooding of the Healey. But the word reached Corbett before the mob.

Corbett was a courageous man. He seized a rifle for self-defense, and sent his informant on the run to tell Heinze that he must stop the mob or there would be serious results. The facts were that Corbett had sent a warning to the men to get out of the Healey before the flood reached them, and Heinze himself had sounded the danger signal in the mine and superintended the escape. Not one man had been trapped.

When the incident threatened murder and riot, Heinze personally informed the mob that they had been misinformed, that no one had been hurt, and that they should disperse quietly to their homes. They believed him, and followed orders. Serious bloodshed was averted, but both sides had been impressed by the demonstration of potential disaster in the dynamite with which they were playing. The very next morning they met and concluded an armistice—as far as underground fighting in the Minnie Healey was concerned.

It was then that Judge Clancy, in the second trial of the Minnie Healey title case, returned a decision upholding Heinze on every point. But, as usual, the Amalgamated again appealed to the State Supreme Court. As far as the company's eminent, able and expensive counsel were concerned, that was merely an incident. They had long been convinced that any case in which they appeared as counsel before Judge Clancy's bench was already decided against them. They were never wrong in that.

Men were employed, as Sam Barker was employed for five years underground, making surveys and arranging data for presentation to Clancy's court with the foregone conclusion that his decision against them had been written by Heinze lawyers even before the trial opened. Eminent experts at a cost of many thousands of dollars were brought in to testify as to the relative positions and apexes of the veins and ore bodies, with the belief that Clancy had already been convinced to the contrary by Heinze's graphs and charts.

Models of mine interiors and geological formations which cost as much as twenty-five thousand dollars each were presented in evidence and brushed aside by Judge Clancy. Underground workings aggregating miles in extent, developed to follow and prove geological theories at a cost of hundreds of thousands of dollars, all told, had been put through to provide data for these mine models. They made no impression upon Judge Clancy, although they ferreted out the geological secrets of Butte hill as no mining secrets had ever been disclosed in the thousands of years of development of the mining industry.

So the Amalgamated attorneys, laboriously, through days and weeks of trial, jammed their evidence into the record knowing that it would be largely ignored in Judge Clancy's decision, but might be used more effectively in the inevitable appeal. Even this hope, however, was upset in the Minnie Healey case. It was not an apex case. It was a suit originally brought by Heinze to establish his title to Miles Finlen's mine over Miles Finlen's repudiation of the alleged oral agreement. When the Supreme Court's decision on the appeal was entered, it was another and most important victory for Heinze.

In the meantime warfare was going on upon a dozen salients. Never in all the checkered and scandalous history of Standard Oil had its potentates met with such lèse-majesté as that impudently offered by F. Augustus Heinze. Never had they had their own methods so turned to their discomfiture. Never had they found an insolent upstart so hard to crush.

It reached a point where Henry H. Rogers sent for Heinze and made him an offer, for the "honest worth" of his properties, doubled to eliminate him as a thorn in the flesh of Standard Oil. Heinze refused. Rogers said he would have no more dealings with the privateer. But Heinze brought more suits, mined more Amalgamated ore and twisted the thorn. Rogers called him in again, and found that his price had gone up to three million dollars and an official position in the Amalgamated to demonstrate to the world that he had won. That was too much for Rogers' pride.

The *Wall Street Journal*, in January, 1904, quoted Rogers as saying: "The flag has never been lowered at 26 Broadway (the Standard Oil fortress), and I'll drive Heinze out of Montana if it takes ten millions to do it."

It would have been cheap at that.

The Amalgamated spent millions in its suits, in Montana politics, newspaper control, propaganda and allied activities. According to an article in Heinze's *Reveille* in September, 1902, the newspaper battle lines were ranged as follows:

		Anaconda Standard
		Butte Inter-Mountain
		Dillon Tribune
		Great Falls Leader
		24 lesser Republican papers
		Butte Miner
The *Reveille*	vs.	*Virginia City Madisonian*
		Missoula Democrat
		Bozeman Chronicle
		Dillon Examiner
		Great Falls Tribune
		Helena Independent
		6 lesser Democratic papers

That, to an interested world including stock-market operators in New York and Boston seemed somewhat in the nature of an unsupported Harvard fullback facing the combined teams of Yale and Princeton. Still, Heinze seemed to be advancing the ball as often as the Amalgamated. It was unbelievable, but it was true. Only the fact that the plays on both sides continued to be trick plays could account for it.

When Thomas W. Lawson, publicity broker par excellence, sent news dispatches to the forty-one newspapers of Montana which spoke for the Amalgamated, asserting that Heinze secretly had sold out, leaving his political friends and mining friends to hold the bag, he supported the charge with an assertion that Heinze's stock in the Montana Ore Purchasing Company, his chief holding company, was actually in the safe of a broker in New York whom Lawson had just seen. All the Butte papers published the story. And the next morning Heinze's stock certificates were displayed in the plate-glass window of his only supporting bank in Butte, piled high for all to see.

An article published shortly afterward in a magazine of national circulation declared that "Heinze's agent had stepped off the north-coast limited train the evening before with the stock. Heinze had simply circumvented Lawson. Lawson was moaning like a mortally injured man over the crimes of Amalgamated and 'the System,' and yet here was a man, the greatest general that had ever led a field against 'the System,' a worthy comrade in any battle, and Lawson was trying to decry and belittle him."

With forty-one newspapers against Heinze's one, and controlling approximately nine-tenths of the jobs in Butte and Anaconda, not to mention those in the great copper fabricating industry of Great Falls and the great logging and lumber camps of the state, the Amalgamated exercised a form of power which even Heinze's elusive dodging could not escape. A trick play was called for.

When announcement had been made that plans had been perfected for the Amalgamated to absorb the Boston and Montana and Butte and Boston properties, John MacGinniss, presumably acting in the interests of Heinze, had purchased one hundred shares of Boston and Montana stock. This represented only one-fifteenth of one percent of the securities. The Amalgamated held ninety-eight percent. MacGinniss sued to have the transfer of Boston stock to the Amalgamated annulled and a receiver appointed, although the Boston and Montana was earning dividends at the rate of more than one million dollars a year.

Judge Clancy granted an injunction to prevent the official transfer of the stock to the Amalgamated, and appointed a receiver as requested. This was the case mentioned in an earlier chapter in which Thomas Hinds, a Heinze man, was awarded two hundred thousand dollars receivership fees for five days' service, while his two "assistants" were awarded thirty-one thousand dollars each and attorneys fifty thousand dollars—all charged to the Boston and Montana.

MacGinniss based his suit on a rule of common law that one corporation could not own, vote or control shares of stock in another. That was good enough law for Clancy. It was not good enough for the opposition. They promptly pushed through a venal legislature a law described in an earlier chapter and denounced by Heinze as "A Bill to Turn the Government of the State of Montana Over to the Standard Oil Company." That was the law which was passed over the veto of Governor Robert B. Smith, who described it as having been sponsored by Standard Oil and the Rothschilds with the purpose of forming a copper trust with permission to confiscate private property.

But before the Amalgamated's grab was "legalized" Heinze had taken a two-hundred-and-eighty-one-thousand-dollar fall out of Boston and Montana. When the stock was transferred under the new law, MacGinniss obtained from Judge Clancy another injunction restraining the Boston and Montana from paying dividends on ninety-eight percent of its stock to its new owners, the Amalgamated. For more than two years the Amalgamated fought in vain to set aside the injunction while two million five hundred thousand dollars in dividends piled up in the Boston and Montana treasury, held there by the action of one man who owned only one-fifteenth of one percent of its stock, and who meanwhile was collecting his own dividends regularly. Then, in October, 1903, on the same day that he decided the Minnie Healey case for the second time in Heinze's favor, Judge Clancy made the injunction permanent.

That was too much for the Amalgamated. It seemed clear that they could not win in the courts or in hand-to-hand underground warfare against a general of such Napoleonic qualities and Al Caponic methods as F. Augustus Heinze. Standard Oil magnates were encountering Standard Oil methods with a vengeance. But Standard Oil had been winning its battles for its own conception of the advancement of civilization for many years. Its directors in the Amalgamated knew more ways of killing a cat than by choking it to death with warm milk. They knew more ways of crushing an opponent than by inducing complacent legislatures to pass favorable laws, suing under those laws, and manipulating the stock market. Standard Oil had always won. It must win again. The Amalgamated tried a new trick.

When Clancy's permanent injunction was announced the company promptly ordered all the mines, mills and smelters of their subsidiary companies to shut down at once. Twenty thousand men employed in Butte, Anaconda, Great Falls, and in the lumber concessions of Montana were thrown out of employment by that order. They constituted perhaps four-fifths of the wage earners in the state. Starvation threatened, almost overnight. No benevolent federal government was prepared, as in 1934, to feed, clothe and shelter the jobless miners.

The company asserted that Clancy's permanent injunction was the culmination of a campaign of years to force the Amalgamated into the position of an outlaw in Montana. It asserted that one man, John MacGinniss, Heinze's right-hand man, had, with the connivance of the court, used one-fifteenth of one percent of the Boston and Montana stock to rob the holding company of the other ninety-eight percent. In that situation, it maintained, its corporate life was threatened and it could not continue to operate until the issue was determined by the Supreme Court.

Heinze's *Reveille* countered by accusing the company of shutting down to intimidate the people and to demonstrate to the legislature and the Supreme Court that it could paralyze the industry and prosperity of the state. "It is doubtless true," said the *Reveille,* "that the motives given out in their signed statements to the public were not the only ones which found place in the minds of H. H. Rogers and his associates."

The press of the world took notice. The *London Times* announced: "It is believed in New York that the Amalgamated Copper Company will benefit by the Montana shut-down, as it will be able to sell on a rising market the large surplus of copper it has accumulated."

Copper market experts asserted that the Amalgamated had accumulated more than one hundred and fifty million pounds of copper by holding it off the market to maintain the price at seventeen cents. With one-third of the world's production at that time cut off in Butte, the market should call for that copper at a higher price. In fact, it did. Copper immediately advanced one cent in London. At that rate the shut-down would prove an economic as well as a tactical achievement for the trust.

The *Boston Beacon* announced: "Montana copper kings are at war. The Amalgamated company, in retaliation for an adverse decision by one of the state courts, has closed down its mines and mills. Twenty thousand men have been thrown out of employment. The effect of this act is to bring home to the body of the people, their dependence upon the good will of the trust."

The *San Francisco Examiner* said: "Seventy-five million dollars were made today by the Rockefeller group of capitalists in the manipulation of Amalgamated copper stock."

The Boston News Bureau sent out a dispatch: "Is the Amalgamated Copper Company appealing for justice? Not for a minute. They have said to the courts and to every political and business organization in the state, 'Deliver the young man

Heinze into our hands or we will destroy both you and him.' The Standard Oil policy of 'rule or ruin' has now come up for judgment."

But all that filled no empty stomachs among the twenty thousand men who had been thrown out of work. Wives and children were hungry. Stoves were cold, and winter threatened. Heinze's assertion that it was all a plot of the trust warmed neither hearts nor hands. Heinze's reiterated claim that he alone had prevented the Amalgamated from reducing wages to two dollars and a half a day fell upon red and angry ears. Now the Amalgamated had reduced wages to zero.

Senator Clark and other locally respected spokesmen for the Amalgamated told the hungry angry men that their jobs had been taken from them at the beginning of winter because of Heinze's virtual blackmailing of the Amalgamated through his control of a venal judge and one-fifteenth of one percent of a subsidiary stock.

Heinze's assumed position of the Moses pretending to lead the miners out of Egypt, out of bondage under the Pharoah of Standard Oil, was derided. His pillar of fire by night and his pillar of cloud by day dissolved with the destruction of a payroll of one million dollars a week which had provided the manna to sustain the exodus.

F. Augustus Heinze was in the tightest spot of his sensational career in Butte. It may be presumed that Henry H. Rogers, William Rockefeller and their associates sat back with a sigh of relief and the grin of a Cheshire cat. After several years of the most vicious battling they seemed to have put the young man precisely where they wanted him.

Their agents approached the president of the miners' union and informed him that the whole trouble was due to Heinze's control of one hundred shares of Butte and Boston stock out of that company's one hundred and fifty thousand capital shares. Let the union buy those hundred shares, agree to the dismissal of the injunction issued by Judge Clancy, and the mines, mills and allied industries would be reopened.

Senator Clark, President Ryan of the Daly Bank and Trust Company, and President Davis of the First National Bank of Butte authorized the union to pay MacGinniss fifty thousand dollars for the one hundred shares, or thirty thousand dollars more than the market price of the stock. The word was carried quickly to Heinze.

Officers of his Montana Ore Purchasing Company, including MacGinniss, vice-president, went into a huddle. When the union's committee waited upon Heinze with their proposition, MacGinniss and the stock were gone. But President Long and his union followers were not to be put off by that technicality. They knew Heinze was the boss. They demanded a "yes" or "no" to their offer. Heinze answered that he would make his reply on the steps of the court-house at four o'clock the next afternoon.

He spread thirty thousand dodgers throughout Butte and Anaconda announcing that he would speak at the hour set. When he appeared ten thousand men crowded the steps, the street, and the window ledges of every building within earshot. President Long of the union and his committee waited upon the steps to renew their offer. Heinze evaded an answer and shouldered through the group to a commanding position, facing the great sullen crowd of men who had for years been looking upon him as their friend and defender and now abruptly changed to regard him as their betrayer.

Now, hungry, they regarded him with enmity and doubt. Was he their Savior or was he Judas? Long again demanded a "yes" or "no."

"I am here to answer to my friends, not to be coerced or blackmailed by Standard Oil," Heinze shouted above the rumble of voices; and looked, clear-eyed, upon the throng.

"We want no answer but one. Will you sell or will you not?"

"That's it," the committee and some in the crowd chorused.

"I'll answer to my friends, all of my friends, ten thousand of them here," said Heinze, cool, imperturbable, convincing.

"Will you sell or will you let these ten thousand men and their families starve?" Long repeated.

"Let him talk," the crowd roared back, shouting down their own authorized spokesman to hear the man who had so long held their partisan loyalty. Heinze had his chance, and for an hour and a half he held that cold, hungry and angry crowd spellbound. Seldom has there been a more dramatic and illuminating illustration of the power of personality over a potential mob.

Shorn of its oratorical trimmings, its appeal to the emotions and prejudices of the miners, its vitriolic attack upon the Standard Oil interests as enslavers of the men who stood before him, Heinze's address may be boiled down to this:

> A brief recital of the fact that John MacGinniss had purchased one hundred shares of Boston and Montana stock at two hundred dollars a share in the spring of 1898, before the Amalgamated was ever organized. That MacGinniss and Lamm had bought a small block of stock in the Parrott Company, subsidiary of the Boston and Montana, at thirty-five dollars a share. That the miners' union, using money provided by their treacherous enemy, the Amalgamated Copper Company, child of Standard Oil, was now offering thirty thousand dollars more for the Boston and Montana stock than any other shares of

that stock were worth on the market. That the union was offering three times the purchase price of the Parrott stock. All that being clear, said Heinze, he was now asked to say whether MacGinniss and Lamm would sell out to the Amalgamated, which would then be in undisputed control of Butte and all Montana—in position to cut wages and starve the miners into a subjection as complete as that of the Roman slaves who were chained to the rocks upon which they labored.

Well, he would answer. Although these ten thousand men had turned against him, he would not turn against them. He would do better than they asked. MacGinniss and Lamm were willing to sell, to sacrifice their interests and the interests of Heinze and the Montana Ore Purchasing Company for peace, and at a price far less than the five hundred dollars a share offered for the Boston and Montana stock. That price, they knew, had been provided by the Amalgamated. They wanted no bribe of the dirty money of Standard Oil. They would sell at the price they had paid plus eight percent interest on their money for the period it had been tied up in these stocks, *provided*——

Ah, here was a joker which revealed the intricate depths of F. Augustus Heinze's shrewdness and power of appeal to the common man——

Provided, that Rogers and Scallon enter a pledge to the miners' union that the Amalgamated mines would be kept in continuous operation through the next year, and the present scale of wages be maintained for at least three years.

And *provided* further, that judgments in the court actions of MacGinniss and associates against the Boston and Montana Company, and the costs of the various actions with reference to this stock be paid.

PHOTOS COURTESY OF MONTANA HISTORICAL SOCIETY, HELENA

Right: *Shy, handsome, charming, and litigious magnate F. Augustus Heinze, who declared, "My fight against Standard Oil is your fight."*

Above: *Converters at Heinze's Montana Ore Purchasing smelter.*

Below: *Butte miners, pawns in the litigations. (Photo of Mountain View Mine, 1,400-foot Station.)*

Montana Historical Society, Helena

The corner of Park and Main circa 1890–1895, looking north toward the "richest hill on earth."

And *provided* still further (another and even more important joker), that the five-thirty-sixths of the Nipper lode claim owned by the Anaconda Copper Mining Company or the Amalgamated Copper Company shall be sold and conveyed to me at the price paid by the Anaconda plus eight percent and the Anaconda and Parrott shall give a deed assuring title to all veins within the Nipper lode claim, and the whole controversy with reference to the Nipper shall be settled to enable me to work it.

If neither the miners' union nor the Amalgamated Company were satisfied with an offer so fair and square to all concerned, Heinze professed himself as willing to submit the entire controversy to a board of arbitration composed of five members, two to be selected by him, two by the company, and the fifth by conference of the other four. He pledged his adherence to any decision reached by such a board.

For the moment he appeared to have won his case. Ten thousand men who had come thirsting for his blood, stood for two hours in the cold of the open street, subject to his hypnotizing personality, and departed in peace, convinced that Heinze was as Christ and the Standard Oil Company as his crucifiers.

CHAPTER XX

MONTANA BOWS TO THE TRUST

THE ten thousand men who had listened to Heinze's hypnotic appeal returned to their homes in a new glow of loyalty to their hero and a new fever of hatred of the trust. But neither the glow nor the fever warmed their wives and children. They awakened, colder and hungrier than ever.

When William Scallon, resident director of the Amalgamated in Montana, proceeded to answer Heinze's specious offers and arguments, the affected miners were cold enough to give heed. With the convincing logic of an able lawyer, Scallon dissected and analyzed each detail of Heinze's emotional appeal.

Heinze's offer to sell the stock in question at cost plus eight percent, said Scallon, was a characteristic Heinze fake. Heinze had said that MacGinniss would sell at that cost, "provided, (1) the Amalgamated pay all court costs and charges of the various actions; (2) the Amalgamated turn over all its rights in the Nipper to Heinze at original cost; (3) the Amalgamated pledge Itself to maintain wages for one year and operations for three years."

Scallon branded these "provideds" as jokers, not to say jokes. The first, he said, would force the company to pay, not the fifty thousand dollars which had been offered for stock which had cost only twenty thousand dollars, but in reality more than three hundred thousand dollars. The added costs were hidden in Heinze's joker providing that in addition to the original twenty thousand dollars cost of the stock the company pay eight percent on that sum for four years, plus two hundred and eighty-one thousand dollars cost of the receivership, plus all court and counsel fees.

Concerning the proviso with reference to the Nipper, Scallon asserted that Heinze had testified in the Pennsylvania case in 1900 that his thirty-one thirty-sixths interest in that mine was worth five million dollars. At that valuation, the share which Heinze was now asking the company to turn over to him as a condition of selling the one hundred shares of Boston and Montana stock was worth eight hundred and six thousand four hundred and fifty dollars. With clear title to the Nipper he could also claim a large part of the Parrott and Neversweat ores. In other words, Heinze's "provideds" called for considerably more than one million dollars profit to him on a twenty thousand dollar investment.

As for Heinze's demand that the Amalgamated pledge maintenance of work and wages, Scallon merely declared that "neither Mr. Rogers nor myself would stand for any cut in wages."

Heinze's arbitration proposal, Scallon branded as an unadulterated fake and appeal to prejudices because, he said, Heinze knew when he made the suggestion that "the Montana law exempts any question of title to real property in fee or for life from arbitration." Heinze therefore could snap his fingers at the decision of the arbiters if that decision were unfavorable to him.

"Mr. Heinze speaks of injunctions against him," Scallon's printed statement concluded. "He and his associates have brought themselves or caused to be brought against these companies in different jurisdictions and states over seventy suits, which is more by fifty percent than all the companies combined have brought against Heinze and his people. Mr. Heinze and his associates have brought thirty-four injunction suits, which is fifty percent, or more, over the number brought by all the other companies combined against him and his associates. . . ."

The allegiance of the miners to Heinze which had been regained temporarily by his dramatic appeal from the courthouse steps began to wane once more. Scallon's refusal on behalf of the Amalgamated to accept Heinze's "offers" seemed sufficient to reasonable men. Heinze's grip on the workers was seriously weakened by Scallon's reply. It was further weakened by a statement published over the signature of Thomas W. Lawson, the Boston broker who had been instrumental in flotation of the Amalgamated and who had a vast following among investors and speculators. Lawson declared that Heinze had repeatedly and secretly negotiated with him and with H. H. Rogers concerning a sell-out to the trust. Heinze's plea of unqualified allegiance to the cause of the miners, Lawson said, was a hypocritical pose. All he wanted was an exorbitant price for his holdings.

Heinze publicly denied the statement and repeated that he would stand by the men in defense of their wages and working conditions as long as they stood by him. But doubt of their leader's loyalty had entered the minds of the miners. There was no doubt that whatever his purposes might be, Heinze's actions had deprived them of their jobs by moving the Amalgamated to shut down the mines.

Business throughout the state was virtually paralyzed. Twenty thousand men and their families were hungry and cold. Others, indirectly dependent upon the Amalgamated payroll, were pinched. Petitions began to pour in to Governor Toole to call a special session of the legislature and pass the laws required to settle the controversy and put the men back to work.

Now indeed the copper trust controlled by officials of the oil trust was in the strategic position which it had long sought. The issue became one of state policy, and of national moment. Could a vast business organization, by starving perhaps one hundred thousand persons, men, women and children, coerce a great state into giving it laws to strengthen its power? That question centered the attention of thinking people throughout the United States upon Montana. President Theodore Roosevelt had already made these people trust-conscious.

Governor Toole was "on the spot." Should he call the special session under coercion or should he let his people starve? He wrote to Mr. Scallon, asserting that he was convinced that it were better for the State of Montana to starve than that "a precedent should be established by which legislation of a given kind should be made a condition precedent to the doing or omission of any act by a corporation or individual of this state."

That doubtless was a noble sentiment, but the Governor realized that it fed no hungry people. And having thus taken the high ground, the Governor proceeded to dicker. Would Mr. Scallon assure him that in the event of a special session being called the Amalgamated would resume work at once, without awaiting results of that session? Mr. Scallon would.

The session was called. The Amalgamated resumed work. The legislature passed the required laws, dictated by the magnates of Standard Oil who were also the magnates of Amalgamated Copper. Every intelligent, literate man in the United States knew what that meant. Every newspaper editor in the country with an eye on national issues and any love for American principles of free government recorded and commented upon that story. Only the pragmatists, or those convinced that the growing civilization and higher welfare of the nation had been brought about and would be furthered by big business, failed to report the incident as the coercion of a state.

So small a newspaper as the *Idaho State Tribune* summed up the case in words typical of the millions of observers who had not been directly involved. "It took the Amalgamated Copper Company just three weeks to coerce Montana into falling on her knees with promises of anything that big corporation might want."

In justice and fairness it should be noted that what the corporation wanted was eminently proper. It had been blocked and defied and thwarted for years by Heinze's legal tactics, chiefly in Judge Clancy's court. It wanted what it termed a "Fair Trial" bill, under which either party to any suit might demand and obtain a change of venue if it had reason to suspect prejudice on the part of the trial judge before whom the action had been set. Nothing could be fairer than that. Indeed Montana was far behind most states in not already having such a statute.

But the fact remained that a combination of corporations had, through the threatened starvation of one hundred thousand persons, forced the unwilling governor of a sovereign state to call a legislature to enact laws for its benefit. That the laws themselves happened to be just was ethically beside the point. A combination of capital, for the first time in the history of the United States, had openly dictated to a state in which it had invested its money and grown vastly rich, and under the laws of which it was operating. The copper magnates were jubilant. One was widely quoted:

"We have demonstrated that the 70,000 voters of Montana who are dependent upon the Amalgamated for a living can be made to think through their stomachs better than through their brains. We have solved the Heinze problem by the only reasonable method."

That was a slight error. The "Heinze problem" was not yet solved. Men of Montana who believed that the fundamental principles of democratic government had been dragged in the muck of Standard Oil gathered in Helena as soon as the special session was called, and proceeded to organize the "Anti-Trust Party," presumably to fight for their state's independence of corporation control and coercion. Two erstwhile governors, several ex-congressmen, and other prominent representatives from both the Republican and Democratic parties were among the six hundred delegates. There were miners, ranchers, lumber men and business men.

Naturally, the Amalgamated newspapers throughout the state promptly branded the organization as a Heinze party, controlled by Heinze's recently organized United Copper Company, "with a capital stock of $80,000,000, and assets consisting of a number of lawsuits and some district judges and county officials in Silver Bow County." But that did not destroy the evidence that thoughtful and intelligent men in Montana were seriously alarmed by the recent proof of the Amalgamated's power to starve the state and coerce its government.

The party did not go far, but it gave Heinze a new weapon and caused new qualms amid the smug satisfaction of the Amalgamated. The justice of the new law could not be disputed although the methods used to obtain it were a disgrace to the civilization to which Montana and the United States had advanced by 1903. The law appeared to disarm Heinze for a time, but when he lost the Silver Bow judges a year later it became a weapon in his hands.

In the meantime some of Heinze's legal actions had passed from the Montana courts to higher tribunals. In the same month that the State Supreme Court had handed down a decision sustaining Clancy's decision for Heinze in the Pennsylvania case (February, 1903), the Supreme Court of the United States upheld Heinze's claim to extra-lateral rights in the Rarus Mine, which had been in litigation for years. If, as the Amalgamated charged, the assets of Heinze's United Copper Company "consisted largely of a number of lawsuits," they were at least proving to be good lawsuits in some instances.

In Montana the copper trust was in the saddle, prepared with whip and spur to ride down F. Augustus Heinze. It controlled ninety percent of the press of the state. It controlled a majority of the legislature. It had proved its power to control a majority of the voters—by starvation if necessary. It had obtained the change of venue law which it had been seeking for years to destroy Heinze's advantage in court actions. It appeared to be getting too rough for William Scallon, its managing director in Montana and head of its legal department there. At least Scallon resigned.

Heinze's *Reveille* announced the withdrawal of his chief personal opponent in Montana in characteristic style. "By deposing William Scallon, Marcus Daly's most intimate friend and associate, from the management of its affairs in Montana, the Standard Oil Copper Trust has severed the last tie that bound it to the old Daly regime."

Writing eight years later, when most of the tumult and the shouting had died, Jerre C. Murphy, in what he called *The Comical History of Montana,* intimated a different reason for Scallon's resignation. "All the power and wealth of Wall Street cannot persuade Mr. William Scallon to become a rascal," he said. More prominent representatives of the day and night life of Butte explained Scallon's resignation as due to the fact that he was no mixer. "He could not take a drink with the boys, play a game of poker with them, slap them on the back, as Heinze or even, to a lesser extent, W. A. Clark could do. He couldn't handle them. He was a good lawyer but he lacked personal charm."

In any event, Scallon was out, and John D. Ryan was in. Ryan had come to Butte as a salesman for the Continental Oil Company when the name of Standard Oil had become too unpopular, and had sold oil to the Anaconda company. Daly trusted him. He became an officer of the Daly Bank in Butte. He was a family man, a church member, and an effective exponent of his own virtues and abilities. When Scallon resigned Ryan appeared to H. H. Rogers to be the logical successor. He got the job. Four years later, after the death of Rogers, he became president.

Murphy's printed opinion of Ryan is in entertaining contrast to his opinion of Scallon. It is also interesting as revealing the boldness of writers of that period. They minced no words. "Organized greed could teach him nothing in selfishness and no awakening conscience disturbed the dreams of his avarice," said Murphy.

Scallon was out and Ryan was in. Montana faced, not precisely a new deal, but another deal.

The Montana Supreme Court reversed Judge Clancy's decision which had made permanent the injunction to prevent the Boston and Montana from paying dividends to the Amalgamated. All the long battle brought in the name of John MacGinniss's famous one hundred shares, the battle which had moved the Amalgamated to throw twenty thousand men out of work and starve the state into submission, seemed to be lost, as far as Heinze was concerned.

But Heinze had another idea. President Theodore Roosevelt's trust-busting activities had produced federal laws which opened a new door to the young man. He invoked one of these laws to carry an action to the New York Supreme Court. There he asked for an injunction to restrain the Boston and Montana, the Parrott and the Anaconda from paying dividends to the Amalgamated. The petition was filed in the name of John MacGinniss. It was virtually a renewal of the famous Boston and Montana fight. The New York Court issued the injunction. The Amalgamated was thrown back to where it had been more than two years earlier when Clancy had taken similar action against it.

Heinze's *Reveille* exulted, and took another crack at the low-lived perfidy, the corruption and the evidence of unscrupulous power exerted by the trust, not only in state but in national affairs. Reporting Federal Judge Hiram Knowles' dismissal of the famous timber suit of the government against the Amalgamated for recovery of five million dollars, the valuation of timber alleged to have been stolen by the trust, the *Reveille* commented:

"Witnesses were hired to leave the country; maps and plats of the denuded sections mysteriously disappeared from the files in which the United States attorney was supposed to keep them, and every effort was made to destroy the smallest vestige of evidence against the company."

A week later the *Reveille* again revealed its uncompromising policy of vicious attack upon the Amalgamated and all its works. A flaring banner line across its front page announced:

H. H. ROGERS STEALS $46,000,000 IN AMALGAMATED DEAL

The story on which the headline was based quoted an article from the *New York Times* of March 31, 1904, reciting Thomas W. Lawson's testimony in the Bay State gas case.

Encouraged by his success in the New York Supreme Court in the injunction proceedings against the Boston and Montana, Heinze then filed suit in a federal court to recover $4,166,166 as his 30/36 share of ore alleged to have been stolen from his Nipper claim by the Parrott company. "It has been an open secret in Butte for years," said the *Reveille,* "that the Amalgamated Copper Company has been stealing Heinze's ore in the most flagrant manner from the Nipper and taking it through the Parrott shaft." The outcome of that suit was announced in a seven-column double headline a year later:

UNITED STATES JUDGE FINES AMALGAMATED OFFICIALS FOR STEALING ORE FROM F. A. HEINZE'S NIPPER MINE

But the story to those who remembered the amount of damages asked went into depressing anti-climax. The value of ore stolen was fixed at $2,236 instead of the $4,166,166 asked in the suit. The fines were $250 each against the Parrott Company and George V. Winchell, and $1 each against John Gillie and Harry Gallwey. The small amount of ore involved tends to substantiate John Gillie's assertion that it was removed in error, not in avarice.

Heinze was still making news which vied in the press of the nation with the Nan Patterson-Caesar Young murder case and the General Slocum disaster in which sixteen hundred persons lost their lives in New York harbor. Underground warfare was resumed. The *Reveille* announced:

WANTON OUTRAGE IN RARUS MINE

AMALGAMATED MEN CAUGHT RED-HANDED DYNAMITING THE STOPES—THEY ARE ARRESTED AND WILL HAVE TO ANSWER TO THE COURTS

"Wednesday afternoon three miners, Thomas Roe, Henry Bill and Thomas Prudhomme, from the Pennsylvania and Amalgamated properties, were arrested for exploding a large quantity of dynamite in the stopes of the Rarus Mine and endangering the lives of several Rarus miners.

"It became advisable for the Amalgamated to destroy the Enargite vein which the Supreme Court decided belonged to the Rarus people. So the explosion followed in order to complicate the litigation. The explosion took place before the men could get away. They were using six-inch fuse and the three were caught on Rarus ground, unable to return to their own workings, with the evidence in their possession . . .

"With an amazing effrontery the muzzled newspapers accused Rarus miners of the explosion. The county attorney has evidence in his possession that will prove that men from the Pennsylvania were responsible."

Two weeks later the *Reveille* found occasion to burst forth again under a flaring headline:

COPPER TRUST AGENTS ATTEMPT TO BLOW UP THE HEALEY SHAFT

"Thursday afternoon the Amalgamated Copper Company, through the workings of its Leonard Mine, attempted to blow up the shaft of the Minnie Healey which was twice adjudged the property of F. A. Heinze in the courts of Montana, and which the Supreme Court of Montana refused to close down by injunction at the request of the Copper Trust. . . .

"For several hours all the men employed in the mine to the number of 500 were imprisoned, as the cages had been hung up before the blasting and several attempts to lower them had proved unsuccessful. There was intense excitement throughout the city when the news spread. At least 1,000 people, many of whom were wives and children of the imprisoned miners, rushed to the scene. . . ."

Murder was in the air. Business could not continue on that basis.

CHAPTER XXI

HEINZE'S LAST COUP

BUTTE was impressed with the advisability of reform. George Wesley Davis, in sketches published by the Cornhill Company of Boston, recites an illuminating incident.

One highly respected old-timer, says Davis, came to him as a trustworthy friend with the request that he witness a wedding. "My eldest daughter is to be married tomorrow," said the old man, "and says the wedding would be a happier one if mama and papa would get married first. So I have arranged to be married early, before the wedding guests arrive. I want you to be one of my witnesses. The public will never suspect for they have never known the bride's name, and a license would mean nothing to them." Such honest procrastination, said Davis, was not uncommon.

The incident was, however, a sign of the times. The wildness and unconventionality of Butte's youth was passing.

Fire, the most disastrous in its history with the exception of the fire and explosion of 1895, scourged the city in the autumn of 1905. The blaze swept from Park Street to Galena, from Renshaw Alley half-way to Main Street. Only a shift of wind after the flames had roared for hours saved the entire city from destruction. Half a hundred business firms were ruined or damaged. The loss of the Symons Company alone was estimated at half a million dollars. It seemed high time for Butte to reform.

Not reform, however, but ambition moved F. Augustus Heinze. The victory of the Amalgamated in starving the state to a point where it had passed the "Fair Trial" bill had been a serious blow to the young "Napoleon." It had started the defection of thousands of his miner friends. He had checked that defection only temporarily by his speech from the court-house steps to ten thousand men. For the first time the rank and file of Butte had become suspicious of his good faith.

In the election which followed they had proved that suspicion by failing to rally to his ticket as they had been rallying for years against the orders of the Amalgamated mines which employed them. Heinze's judges had been swept out of office. Heinze's attempt to continue his legal tactics under the antitrust laws despite changes of venue had been successful in some instances, but by no means as successful as his personal control over Silver Bow courts.

He recognized the handwriting on the wall. He recognized in John D. Ryan, successor to William Scallon as managing director of the Amalgamated in Butte, a man with whom he might profitably do business. His assets, contemptuously cited by the Amalgamated press at the time his United Copper Company was organized as "consisting of a number of lawsuits and some district judges," still included the lawsuits, although they no longer included the judges.

While still continuing his *Reveille's* bitter attacks upon the trust, he entered into secret negotiations with Ryan. But he came on no bended knee. Although the trust might scoff at his lawsuits in the guise of "assets," it had been forced to expend millions of dollars because of and in opposition to those suits. Although it superficially ignored any other assets it knew that Heinze, through his Montana Ore Purchasing Company and his United Copper Company owned or controlled the Minnie Healey, the Rarus, the Nipper, the Belmont, the Hypocka, the Corra-Rock Island, and valuable smelting and milling properties.

Here was something worth trading for. Heinze had made good his threat of seven years before to give A. S. Bigelow of the

Boston and Montana a battle which would be heard of from one end of the country to the other. With scores of suits still in the courts, he might very well continue to do so. John D. Ryan was ready to listen to reason.

He met Heinze by appointment in a private room of a Butte hotel. Charles R. Leonard, Heinze's attorney, and still one of the leading lawyers and most respected citizens of Butte, was called. A proposal of peace was outlined before him. Heinze and Ryan were personally on the friendliest of terms. There seemed nothing in the way of a settlement except a thousand details.

The scene of negotiations was moved to New York, where Ryan could have the benefit of advice from that master of financial strategy, Henry H. Rogers, and where Heinze could have the support of Charles R. Leonard without all Butte knowing what was going on. Offices with widely separated entrances were arranged for continuation of the parley.

If Butte learned that Heinze and Ryan were negotiating, Butte might easily explode. If Wall Street learned of it, there was no telling what effect it might have upon the copper stocks market. So carefully were the negotiations hidden that they continued for seven long months without a word leaking out. A price was put upon each and every item of Heinze's holdings.

John D. Ryan and Thomas F. Cole, who had gained a high place in big business through his organization of Lake Superior iron mines for the Steel Trust, organized the Butte Coalition Company, with the understanding that it should go into the Amalgamated holding company, when the deal was completed. Incidentally that would save the face of Henry H. Rogers. The Standard Oil magnate would not be bowing to the upstart Heinze if he took over control of a new company which happened to have bought Heinze's interests.

Big business men are sometimes strangely sensitive about such matters, even though their consciences appear entirely insensitive to the starvation of thousands and the coercion of states to effect their purposes. Heinze would not be selling out to Standard Oil, which he had cursed and berated for years in his appeals to the common people of Butte.

All was, as the saying went in 1906, "hunky-dory." The so-called Cole-Ryan interests, operating through the new-born and therefore innocent Butte Coalition, paid F. Augustus Heinze, the young man who had entered Butte seventeen years earlier as a surveyor at five dollars a day, ten million five hundred thousand dollars for his mines and lawsuits. One hundred lawsuits involving property worth fifty million dollars were dismissed.

Heinze was rich, free, but not satisfied. The Amalgamated Copper Company now ruled Butte without a rival. But Standard Oil, though it had superficially saved its face, was sore from the beating. The battle of Butte was finished, but there impended a repercussion which would shake the nation.

Charles R. Leonard, who was in at the death, asserts that the nation-shaking event of the following year was in no way due to that soreness of Standard Oil. Perhaps he is right. Such vast corporations, commonly called soulless, conscienceless and heartless, are not supposed to carry and act upon personal grudges. But there is testimony in the record to the contrary.

With ten million five hundred thousand dollars in his pocket giving him some right to consider himself a Napoleon of finance, F. Augustus Heinze, at the age of thirty-seven, came to New York in 1906 prepared to "take" the world. There he established his brothers, Arthur and Otto in the Stock Exchange firm of Otto C. Heinze and Company. He was a financier. A great financier needed a bank. Heinze bought one, the Mercantile National, from Edwin Gould, "paying an enormous price," according to the *Engineering and Mining Journal* which had employed him as a reporter fifteen years earlier.

Heinze was all set to give Standard Oil another trimming. But now he was fighting on their ground, not his own. Still, knowing more about copper than anything else, he planned to continue his rise, at least for the time being, through major operations in that field.

The United Copper Company which he had organized with a capitalization of eighty million dollars several years earlier had disposed of most of its assets to the Butte Coalition in Heinze's settlement. But it was still in existence as a corporation although its capital had never been written down to allow for subtraction of its most valuable holdings. It seemed to Heinze to offer the opportunity he sought. He organized a pool and began to buy.

Whether the crash in stocks which occurred on March 13 and 14, 1907, was in the nature of a warning chivalrously issued by big-hearted operators, has never been discussed. In view of Wall Street's history it seems rather to have been a slightly premature attempt to catch Heinze and his friends out on a limb. If so, it failed. Nevertheless it should have been a warning if Heinze's experience in Wall Street had been equal to his experience underground. Amalgamated Copper lost eighteen points and American Smelting and Refining twenty points in two days.

But Heinze, more directly interested in United Copper, survived. Instead of drawing in his horns, he thrust them out. Had he not "taken" the Amalgamated for ten million five hundred thousand dollars only the year before? He had. Was not the redoubtable Henry H. Rogers, vice-president of Standard Oil, president of the Amalgamated, director of the greatest trusts in the country, in control at that time? He was. Did not Heinze own a bank and a seat on the Stock Exchange? He did. Did he not have a great following of men eager to cooperate under his direction in a copper pool which should rival the thirty-six million flotation profits of Amalgamated? He did.

The business of buying United Copper stock continued. Another flurry in the market in mid-summer failed to check it. As autumn approached word leaked out that there was a corner in United Copper. Quotations leaped in a few days from 37 to 60.

Who should know more about the stock market end of copper than the magnates of Amalgamated, of Standard Oil, who controlled the majority of copper mines and smelters of America? No one. Who could be more familiar with the tricks and treacheries of Wall Street than the men who had made seventy-five million dollars grow in a single day of 1899 where thirty-nine million dollars had grown before? Not F. Augustus Heinze, miner.

The Heinze pool, knowing the precise number of shares they had purchased, knew that there was no corner. What they did not know was that the opposition had promoted the rumor of a corner in order to send the price skyrocketing. Another thing they did not know was that this opposition had been holding United Copper shares, purchased at a low price, all through the early stages of Heinze's bull operations. This was what had limited offerings of the stock on the market and given the Heinze group an entirely false idea of their power and importance when they cooperated to push up the price.

In the rapid soaring of the price from thirty-seven to sixty the opposition skillfully unloaded through various brokers supported by various banks which appeared eager to lend on the stock. The certificates were actually in the hands of the lending banks. The trap was set, and Heinze and his associates were in it. All that was necessary was to pull the trigger. It was pulled with an effective jerk.

The banks—not Heinze banks—called their loans on United Copper. Heinze, his brokers and his associates were forced to sell the stocks held by the pool to meet the called loans. There were no buyers of any importance. This was too big a deal for the pikers. And Heinze had chosen a stock for his coup which had virtually no intrinsic value. The price which had been forced up to sixty slumped off to ten. The groans of the stricken echoed through Wall Street, through Boston and through Butte.

The State Savings Bank of Butte suspended payments. A run began upon Heinze's Mercantile National Bank in New York, his pride and joy, his medal of victory over the Amalgamated in the long battle of Butte. It could not meet its adverse balance at the Clearing House. Neither could the National Bank of North America, the New Amsterdam National Bank, and several smaller institutions.

Now F. Augustus Heinze was on the spot. On bended knee he appealed to the Clearing House Committee for aid. There might have been some small solace to his pride in the fact that James Stillman, president of the National City Bank, "the Standard Oil bank," fiscal agents in the flotation of Amalgamated, had finished his term on the committee in the preceding year, but it was very small solace. The committee included men as cold of eye and heart as Stillman. They had been trained in the same school of finance as Stillman.

They turned the screws on F. Augustus Heinze, and upon the other banks which were in the same position as the Mercantile National. "Resign!" they told Heinze. "Get out! Scram!" or words to that effect, "and we will see what we can do. Otherwise your bank goes down in the wreck with you."

Heinze had never before known when he was beaten. This time he did know. The firm of Otto C. Heinze and Co. failed. F. Augustus Heinze was discredited, ruined. It had taken the subtle and pitiless forces of Henry H. Rogers and his associates just one year to take away the ten million five hundred thousand dollars which they had paid to Heinze after seven years of warfare.

"The bank was saved, but the panic (1907) was started," said an article published in *Current Literature* in the following year.

"It then appeared that the panic was precipitated by the struggle to get rid of Heinze," says the sketch of his life published in the *Dictionary of American Biography* with the perspective of twenty-five years.

"This was the beginning of the panic of 1907," said the *Engineering and Mining Journal.*

"It, the panic, began through the inability of certain persons who were speculating in United Copper Company stock to continue the operations for a rise . . .," said Charles A. Conant's *A History of Modern Banks of Issue.* "These conditions afforded an opportunity for the more conservative bankers who controlled the (Clearing House) committees to accomplish what they had long desired—the elimination of a group of speculators from the banking world."

Despite the contrary opinion of Charles R. Leonard, who was Heinze's chief counsel in the earlier triumphant deal with the Amalgamated, there seems to be considerable and reputable testimony that Wall Street's punishment of Heinze precipitated the panic of 1907. It may not have been the cause but it was the spark in the powder magazine. And there is no doubt that there were few more powerful figures in Wall Street in that autumn of panic than Henry H. Rogers, William Rockefeller, and James Stillman of the National City Bank. Perhaps J. Pierpont Morgan was one. J. Pierpont Morgan was also one of the original directors of Amalgamated Copper.

No individuals, no group of men in the world had more reason than the management of the Amalgamated Copper Company to wish to see Heinze's hide nailed to the door of No. 26 Broadway, home of Standard Oil.

It was nailed.

In the course of the flaying which preceded the nailing, other incidents had developed into a serious situation. The stock-market crash of the spring and the flurry of the summer had left the market in a state of jitters. The crash of the autumn shook the nation. In addition to the fight to "get" Heinze, a battle had developed between the Clearing House and the trust companies, of which the Knickerbocker Trust, with sixty million dollar deposits, was perhaps the greatest.

When the market crashed, a run on the Knickerbocker was precipitated. Eight million dollars were paid out to depositors. The Knickerbocker failed with fifty-two million dollars still owing. There was no stopping the panic after that. Probably nine out of ten persons over the age of forty now in the United States remember the Clearing House certificates which passed for money in the grave days of 1907.

"The forcing of Heinze and Morse out of their chain of banks," said *Current Literature* a few months later, "was the first signal to the public that the trouble had at last reached an acute stage, and the closing of the Knickerbocker Trust doors was the first of the consequences that gave a chill to the nation's prosperity. These things once accomplished, Morgan and the Standard Oil joined forces to keep the fire from spreading any farther than necessary."

How it came out is another story. Its inception only is of interest in this record.

Even Wall Street recognized the fact that a house-cleaning was necessary. In the course of that house-cleaning Charles W. Morse, one of the most notorious bankers of New York, and F. Augustus Heinze were indicted on various charges of fraud. Morse was sent to the federal prison at Atlanta. Heinze was acquitted.

"Butte went wild with joy," said the *Engineering and Mining Journal.*

CHAPTER XXII

ANACONDA LIVES UP TO ITS NAME

IN ORDER to do no one an injustice it might be said that F. Augustus Heinze was to the panic of 1907 as the Archduke Franz Ferdinand was to the World War. And Heinze was of little more use in the subsequent hostilities than the slain Heir Apparent to the throne of Austria-Hungary.

He returned to Butte "triumphantly vindicated" by the federal court which had sent his associate, Morse, to prison, but that vindication was of little more value than the "triumphant vindication" of William A. Clark through his trick appointment to the United States Senate after that body had repudiated his election on the grounds of improper practices.

Heinze's skin had been left on the door of No. 26 Broadway. His heart had been divided among the wolves of Wall Street. Butte was "wild with joy" because it saw only the tailored raiment covering the flayed body, and the familiar smile hiding the torn heart. Heinze had been their hero for many prosperous years. They had repudiated him when his fight with the Amalgamated had led that organization to throw twenty thousand men out of work, but their memories were short. Heinze had stood for high wages, low smelting costs, opportunity for independent mines, prosperity. He had given them thrills, entertainment and money. They expected him to do so again.

But the fighting heart was gone. Most of his material resources were gone. He owned virtually nothing in Butte. The Amalgamated controlled the hill, the smelters, the plants at Great Falls, the timber resources of the state, a large part of the water-power rights, the leading newspapers, the great company store, Hennessey's, the political organizations.

There was not a crack into which Heinze could insert a foot to start a new climb. He had saved the Stewart Mine in the Coeur d'Alenes and the Mascotte tunnel in Utah when he had turned his Butte holdings over to the trust. They were far enough away to hold no interest for the Amalgamated. Even Heinze could not claim the veins of Butte hill had their apex in properties a hundred miles away. They had allowed him to keep these properties as not being worth the price he asked, and as being no possible source of danger to their own purposes.

Unable to gain a new footing in Butte, Heinze turned his attention to these neglected properties. He was still an engineer, although that fact had been forgotten in his last ten years of warfare. He made the mines pay sufficiently to maintain him in the luxury to which he had become accustomed. But the thrill of battle, the zest for life, were gone. The romance of great achievement against great odds was gone. Heinze had burned himself out.

He sought solace in a personal romance at the age of forty-one, marrying Berenice Henderson, an actress, in New York. There was little comfort in that. Even the birth of a child left him cold. In two years came divorce. A year later there was a reported reconciliation, but Mrs. Heinze died soon after. Heinze himself was in bad health, suffering from cirrhosis of the liver. A year later an acute attack ended his life, in Saratoga, New York.

The boldest, most dramatic, and, according to his enemies, the most unscrupulous figure in the history of mining in America was gone. The greatest of his enemies, the Amalgamated Copper Company, lived on. To be sure, under the pressure of laws promoted by the trust-busting President Theodore Roosevelt, it slightly altered the form of its organization and changed its name to the Anaconda Copper Mining Company.

The assets, the personnel, the spirit of the trust remained virtually unchanged. It had proved itself an anaconda. Quite properly it became, and continued to be, The Anaconda. Opportunity was still beating a drum in America to summon those capable of seizing upon advantage.

The directors of the Anaconda still directed the destinies of Butte, of the town of Anaconda, of Great Falls, of Montana, and continued to enlarge their sphere of influence to the economic and political destiny of other regions. The original plan to rule the copper market of the world, and so to control the electric light and power industry of the world was restricted but not destroyed.

Efficiency took the place of romance in Butte. Energy was applied to the mining, smelting, refining and fabricating of copper and by-products instead of legal battles. Stockholders received more dividends, lawyers less. Production increased from fifty million dollars a year to one hundred million dollars in each of the years of the World War. The coils of the Anaconda were encircling more and more property, more and more enterprise.

In the year of Heinze's death it acquired control of the International Smelting and Refining Company. In the second year of the war it threw its coils into South America and seized control of the Andes Copper Mining Company. A year later it took over the Santiago Mining Company. After the temporary depression of 1921 it acquired the tremendous assets and plants of the American Brass Company in Waterbury, Torrington and Ansonia, Connecticut. Before the death of Senator William A. Clark at the age of eighty-six, it purchased his still extensive remaining assets in Montana.

It expanded and improved its refining and fabricating plants at Great Falls. It installed on Butte hill electric hoists at a cost of five million dollars each, capable of raising ores from a depth one thousand feet lower than had yet been reached in the mines. In 1926 it exercised an option on the vast zinc deposits of the von

Giesche properties in Upper Silesia, between East Prussia and Poland.

It became the largest producing-fabricating copper mining enterprise in the world, mining twelve and one-half percent of the world's copper, smelting eighteen percent, refining twenty-two percent, fabricating twenty percent.

Marcus Daly's vision of "the richest hill on earth" was justified. The opportunity which he recognized in the barren hillside of Butte in 1876, and which was seized with varying degrees of success by perhaps one hundred thousand men, produced two billion two hundred million dollars in copper, silver, zinc, lead, gold and other minerals in little more than half a century.

Its production of refined copper would make a solid block of metal the size of the largest office building in the world. Its zinc would form a solid block the size of the world's largest battleship. Its lead would fill a mold as large as the hull of a destroyer. Its silver, refined to sterling, would build a statue like the Statue of Liberty. Perhaps nine-tenths of the home-owners in the United States who take advantage of the New Deal's assistance for repairs and improvements in their kitchens and bathrooms, will use a part of its products.

It would take the average motorist, traveling at a city traffic rate of twenty-five miles an hour, thirty-five hours to drive through its workings under Butte hill.

There is concrete, copper-riveted, brass-bound, silver-trimmed evidence of the accomplishments of corporate efficiency in America.

But who got the money? An average of perhaps one hundred thousand persons have been living on part of it in the form of wages in Montana alone through more than half a century. Probably one hundred million persons in the United States who more or less frequently use electric lights, telephones, radios, automobiles, modern plumbing, waffle irons, toasters, refrigerators and so forth have taken their bit in comfort and luxury through the developments in electric appliances paralleling the production of copper.

Neither could have reached its high point without the other. Without copper, no electricity in commercial form. Without electricity, no such automobiles as we enjoy today, no radios, no telephones, no electric lights, no jobs for a million men now engaged in production, sales and service. Without copper, and its union with zinc, no such modern and sanitary plumbing as half of the United States now takes for granted. More jobs for more hundreds of thousands. More profit in comfort and health and luxury for millions.

And what of the individual? F. Augustus Heinze, a Brooklyn schoolboy, trained as an engineer, wrested ten and one-half million dollars from the greatest trust of the America of his day. In so doing he revealed opportunity to thousands. Some profited directly only to the extent of earning a living. Others of his associates, as late as 1934, twenty years after his death, still tread upon Oriental rugs which must have cost them thousands. Hundreds, helped directly and indirectly by his stimulation of independent mining and his great reduction of smelter costs, have lived out their days in affluence. Heinze died, a young man, forty-five. Who shall say he did not crowd more life into those forty-five years than many a man into seventy?

Marcus Daly, the Irish immigrant, untrained in any school except that of hard work and experience, saw opportunity in a blue-gray rock, and took millions from it in the twenty-five years of life which remained to him. Thirty years later his son-in-law, James W. Gerard, one time United States Ambassador to Germany, presented to the residents of the Bitter Root Valley of Montana the Marcus Daly Memorial Hospital, at Hamilton, in behalf of the widow, Margaret P. Daly, and family.

William Andrews Clark, at the age of twenty-four, entering the wilderness of Montana with an ox-team, weary with hundreds of miles of mountain travel, heard the call of opportunity in a gulch a day's travel away. Without a pause he struggled on. For him there was opportunity in a gravel pit which he could work with his own hands. There was opportunity in a pair of elk horns found beside the trail. There was opportunity in a case of eggs which he might buy at twenty cents a dozen and sell at three dollars. That was the foundation of one of the largest fortunes ever gathered by one man in the history of the United States.

Clark accumulated capital simply by grasping every opportunity and improving upon it. Farm boy, country schoolteacher, miner, merchant, banker, financier, politician, capitalist, patron of the arts, philanthropist. Still, not a man to be admired? Perhaps. Certainly not a man of tender conscience. But just as certainly a vivid exemplification of definite accomplishment in a day when Americans stood upon their own feet, without fear or favor.

That was not so long ago. Clark died in March, 1925, at the age of eighty-six. He had been active in the direction of his vast affairs up to the week of his death, through more than half a century of experience in the management of personally earned millions.

But ignore the millions, temporarily. There is too much talk of millions in the gauging of success. Ignore the bad taste of his fifteen-million-dollar mansion in Fifth Avenue, New York, with its one hundred and twenty-one rooms, its thirty-one baths, its four picture galleries, its fifteenth- and sixteenth-century tapestries, its statuary, French porcelain and imported ceilings. Ignore the egotism which led him to bequeath his art collection to the Metropolitan with the proviso that it be maintained intact in a special wing of that famous gallery—a restriction which moved the trustees to refuse the gift because many of the items were not worthy.

Consider in this day of economic tension, only a decade after his death, but one phase of Clark's vast and diversified interests—

his bank. That bank was founded in Deer Lodge, Montana, in partnership with R. W. Donnell in 1869. A few months later S. E. Larabie joined the firm. In 1872 it obtained a charter as the First National Bank of Deer Lodge. Within a year Clark revealed his reaction toward government regulation. The charter was surrendered. The bank again became a private bank. Clark would stand upon his own feet. When Butte was rising to greatness the branch there was reorganized under the title of W. A. Clark & Brother, Bankers, with every share of capital stock in the hands of W. A. and J. Ross Clark.

In that situation, depending upon nothing but the brains and earnings of its founder-owner-manager, it carried blithely through the panics of 1893 and 1907. Every dollar of Clark's millions earned in the mines and smelters of Butte, in the vast development of the United Verde, in timber and water-power interests, in the building and subsequent sale of the railroad from Salt Lake City to Los Angeles, and in all allied activities, was behind that private bank.

No state or federal laws restricted the operation of Clark's judgment. His personal fortune was ten times the amount of deposits in the bank, and it was all behind those deposits. Even when the United States Senate decided that Clark's political methods disqualified him for a seat in that august body, not one of the thousands of depositors in Clark's bank questioned the safety or the wise and honest management of their funds. "It is doubtful," said the *Northwestern Banker* in an article published in March, 1922, "whether any bank in this country, not excluding the firm of J. P. Morgan and Company, has such immense resources behind it."

That confidence, built upon a man, was never betrayed. When the late Senator Robert LaFollette listed Clark as one of one hundred men who owned America, and cited fourteen great corporations in his hands to prove it, he neglected to recite a more revealing fact.

Differing from the other ninety-nine "owners of America," Clark stood alone, unique, in the fact that of all the great enterprises with which he was connected, not one share of stock nor bond issue by any one of them was either listed or quoted or could be bought on any stock exchange in the United States.

William Andrews Clark earned his money. In his employ hundreds of men reached affluence. Thousands earned a living. At his death approximately one million dollars went to charity and philanthropy. His art collection went to the Corcoran Galleries. The remainder of his millions went to his widow, his sons and his daughters. At the death of his son, W. A. Clark, Jr., in 1934, another great gift came back to the public in the form of the carefully collected and beautifully housed library bequeathed to the University of California at Los Angeles.

William A. Clark was convicted of political practices so outrageous that the United States Senate refused him the seat to which he had been elected. Two years later, he marched down the aisle, and took that seat. Twenty-odd years later the leading men and women of Montana gathered in the rotunda of the capitol at Helena, and unveiled a bronze bas-relief of Clark's shaggy head, and this inscription:

WILLIAM ANDREWS CLARK
PIONEER PROSPECTOR AND MINER
MERCHANT BANKER RAILROAD BUILDER
BENEFACTOR OF CHILDREN AND PHILANTHROPIST
THIS MEMORIAL IS ERECTED BY
THE SOCIETY OF MONTANA AND OTHER FRIENDS
AS A TRIBUTE TO HIS GREAT ACHIEVEMENTS
AND TO PERPETUATE HIS MEMORY

That could happen in America.

INDEX